RE-IMAGINING MILK

The Anthropology of Stuff is part of a new series, *The Routledge Series for Creative Teaching and Learning in Anthropology*, which traces the biographies of commodities, illuminating for students the network of people and activities that create their material world. Each book also helps students understand that they benefit from exposing their own lives and activities to the power of anthropological thought and analysis. Milk is a fascinating food: it is produced by mothers of each mammalian species for consumption by nursing infants of that species, yet many humans drink the milk of another species (mostly cows) and they drink it throughout life. Thus we might expect that this dietary practice has some effects on human biology that are different from other foods. *Re-imagining Milk* considers these, but also puts milk-drinking into a broader historical and cross-cultural context. It asks how dietary policies promoting milk came into being in the U.S., how they intersect with biological variation in milk digestion, how milk consumption is related to child growth, and how milk is currently undergoing globalizing processes that contribute to its status as a normative food for children (using India and China as examples). Informed by both biological and social theory and data, *Re-imagining Milk* provides a biocultural analysis of this complex food and illustrates how a focus on a single commodity can illuminate aspects of human biology and culture.

Andrea S. Wiley is Professor of Anthropology and Director of Human Biology at Indiana University, Bloomington. She has conducted long-term field research in India. Her previous works include *An Ecology of High Altitude Infancy* and *Medical Anthropology: A Biocultural Perspective* (with John Allen).

The Routledge Series for Creative Teaching and Learning in Anthropology
Editor: Richard H. Robbins, SUNY at Plattsburgh

This Series is dedicated to innovative, unconventional ways to connect under-graduate students and their lived concerns about our social world to the power of social science ideas and evidence. Our goal is to help spark social science imaginations and in doing so, new avenues for meaningful thought and action.

Forthcoming

Coffee Culture by Catherine Tucker
Lycra by Kaori O'Connor
Fake Stuff: China and the Rise of Counterfeit Goods by Yi-Chieh Jessica Lin
Reading the iPod as an Anthropological Artifact by Lane DeNicola

RE-IMAGINING MILK

Andrea S. Wiley

Routledge
Taylor & Francis Group

NEW YORK AND LONDON

First published 2011
by Routledge
270 Madison Avenue, New York, NY 10016

Simultaneously published in the UK
by Routledge
2 Park Square, Milton Park, Abingdon, Oxon OX14 4RN

Routledge is an imprint of the Taylor & Francis Group, an informa business

© 2011 Taylor & Francis

The right of Andrea S. Wiley to be identified as author of this work has been asserted by her in accordance with sections 77 and 78 of the Copyright, Designs and Patents Act 1988.

Typeset in ITC New Baskerville by Glyph International Ltd
Printed and bound in the United States of America on acid-free paper by Walsworth Publishing Company, Marceline, MO

Library of Congress Cataloging-in-Publication Data
Wiley, Andrea S., 1962–
Re-imagining milk / Andrea S. Wiley.
  p. cm. – (The Routledge series for creative teaching and learning in anthropology)
Includes bibliographical references and index.
1. Milk–Social aspects. 2. Milk–Social aspects–United States. 3. Milk consumption. 4. Milk as food. I. Title.
GT2920.M55W55 2009
394.1'2–dc22   2010022294

ISBN 13: 978-0-415-80656-5 (hbk)
ISBN 13: 978-0-415-80657-2 (pbk)
ISBN 13: 978-0-203-83697-2 (ebk)

CONTENTS

PREFACE

I have long been fascinated by milk. Ever since I was introduced to human variation in adult lactase production in an undergraduate course in biological anthropology, I have been intrigued by the significance of this substance as a human food. Having never questioned the normalcy of cows' milk on the dinner table or my morning cereal, the recognition that not all humans drank milk—indeed that many would become sick from drinking it—led me to start asking questions about this topic. I soon discovered that many people had strong feelings about milk, which ranged from very positive to overwhelmingly negative, and that these were accompanied by an equally wide array of consumption patterns. I wondered why drinking the milk from cows was viewed in the United States as "normal" while the very sight of mothers feeding their infants breast-milk provoked embarrassment or even charges of indecency.

After getting my PhD in anthropology and working on other projects related to maternal and child health and nutrition in India, I returned to my longstanding interest in milk (which is very important in the Indian context, as you will see in Chapter 5) and began work on how anthropological knowledge about human biological variation in lactose digestion articulated with dietary policies that explicated promoted—indeed mandated—milk for daily consumption by all. I also began to think about the distinctiveness of cows' milk as a human food: milk is, after all, the only mammalian food produced in order to be consumed, its target consumers are calves rather than human children or adults, and it contains a variety of nutrients and bioactive molecules that support the growth and development of calves. Might lifelong consumption of such a substance have some unusual or unexpected effects on human biology? How did such a food ever come to be considered, at least in some societies, an expected part of human diets?

In this book I take up some of the threads of my interest in milk and weave them together in an effort to do three things: (1) uncover how cultural forces shape our views about the "naturalness" of cows' milk consumption and how these views have been molded across evolutionary and historical time scales; (2) provide a brief overview of what is known about the relationship between milk consumption and some aspects of human biology such as child growth; and (3) to show how a holistic investigation of a particular food illuminates some of the complexities of human culture and biology. Anthropological analyses of single foods have been rare (a notable exception being Sidney Mintz's brilliant work, *Sweetness and Power* (1985)), and until relatively recently not many anthropologists have regarded food as a serious topic of scholarly inquiry.

I start with an exploration of milk as a biological substance and outline the biocultural perspective I will use throughout the book. This means that I approach milk as both a biological substance with specific physiological effects on the body, and as a food that is part of complex global and local agricultural economies with an array of cultural meanings across and within societies. This is followed by an examination of human variation in lactose digestion, how it has been reconciled with dietary and health policies that strongly encourage milk consumption in the U.S., and the kinds of rhetoric that have been deployed in the dialog surrounding the value of milk in the diet of Americans. In this discussion I develop the term *ethno-biocentrism*—an elaboration of the more familiar anthropological term *ethnocentrism* to include the interpretation of other people's bodies as well as behavior only in relation to those of one's own body and culture, generally with the view that one's own is "better" than the other, or that one's own is "normal" and others are somehow "abnormal."

From there I turn to the history of fresh milk consumption, considering how this can be considered a very "modern" phenomenon whose salience has waned in many countries, while it has simultaneously taken on new positive meanings as milk-drinking has achieved global normativity. In Chapter 4 I review the scientific data relating milk consumption to various biological outcomes, with a focus on child growth: Does milk make children grow? This is contextualized within a broader discussion of assumptions made about milk and current debates about the "goodness" of milk in relation to a number of biological outcomes. The way in which "natural" connections between milk and child growth have been deployed to expand milk consumption to a global scale is covered in Chapter 5, which uses India, China, and the U.S. as examples. The book closes with a consideration of the controversies surrounding milk and why milk continues to generate vigorous and often rancorous debate.

This book is by no means a complete analysis of milk or milk-drinking around the world. Indeed, I have spent little time on milk production, a rich and fascinating topic, focusing instead on milk consumption, particularly, although not exclusively, in the United States. Highlighting consumption allows me to integrate biological and cultural analyses, to see what forces shape individual ingestion practices, and how these have biological effects. I also consider how these have changed over time, and how milk is positioned vis-à-vis desires for different aspects of biological and social well-being. I hope to provide the reader with a critical but also even-handed analysis of this particular food, and how it reveals much about our assumptions about milk's presence in a "normal" or "healthy" diet.

ACKNOWLEDGMENTS

I would like to thank Richard Robbins for his interest in my work, and his vision of it in relation to the new Routledge "Anthropology of Stuff" book series. At Routledge, Steve Rutter and Leah Babb-Rosenfeld were wonderful to work with, and thanks to Sarah Stone, Victoria Brown, and Kate Legon.

I greatly appreciate the comments from reviewers along the way: Crystal Patil at the University of Illinois Chicago, Mark Jenike at Lawrence University, and Deborah Crooks at the University of Kentucky.

I also owe an intellectual debt of gratitude to Melanie DuPuis, whose book, *Nature's Perfect Food* (NYU Press, 2002) was inspirational, and provided the historical research that was essential to the framing of my questions regarding the normative relationships between children and milk consumption. I have leaned on her astute and thorough analysis in many ways throughout this book. I take full responsibility for any remaining shortcomings of the book.

I received funding from Indiana University for a research trip to India to begin my explorations of milk there, and from the Wenner-Gren Foundation for Anthropological Research to carry out my current research on the relationship between milk consumption and child growth in Pune India.

I am also indebted to my parents, Roy and Marilyn Wiley, for their interest in this topic and my mother's longstanding efforts to provide schoolchildren with wholesome, nutritious, and delicious meals (which also happened to include — by mandate—milk). My sister, Marcia Wiley, continually supplied me with popular media materials about milk and never failed to encourage me. Finally, this book would not have been possible without the love and support of my husband Richard Lippke, and my two sons Aidan and Emil, both of whom have complicated relationships with milk.

1

INTRODUCTION

On the "Specialness" of Milk

Growing up in the United States most of us have been told: "Drink your milk!" If we stubbornly refuse or ask why, we are admonished that drinking milk will "help us grow" or "grow strong bones." These are ubiquitous messages in the U.S. and many other countries, especially those with well-established dairy industries, and increasingly in countries where milk traditionally has not been produced or consumed. Milk is endorsed by the U.S. government, appearing in the Dietary Guidelines for Americans and its more familiar visual representation the Food Pyramid, which recommends that all Americans consume two to three servings of milk or dairy products each day. It appears as a strange white mustache on celebrities in print advertisements and on lunch trays at school. It probably was featured on posters in your elementary classroom and in whatever nutrition education you received in school. Perhaps you drink multiple glasses of milk daily, or despise the taste and avoid it like the plague, but regardless of their actual milk consumption practices, most Americans "know" that milk is something they "should" consume on a daily basis.

Where did this idea, and other ideas related to milk come from? What is milk, and why is it considered such an important food in the U.S. and in many other countries? In this book we will be concerned with these questions and several others. Why do some people get sick when they drink milk while others experience no problems whatsoever? What happens if you don't "drink your milk?" Does milk help you grow? What role does milk play in the diet and economies of societies around the world? What is the cultural significance of milk in the U.S.? Is milk "special" or "unique" in ways that set it apart from other foods?

In this book I will explore the diverse meanings of milk and the factors that influence milk consumption. The emphasis will be on the United States, but I will also consider how milk is understood and consumed in other contexts. Unless otherwise noted, "milk" will refer specifically to cows' milk, which is the type of milk most widely produced and consumed in the world today.[1] Milk is somewhat different from other foods in your diet. First, it is one of only two animal foods (the other being honey) that is produced by individuals of a particular species to be consumed by members of the same species. Second, unlike honey, which is consumed by all members of the beehive, milk is consumed only by infant mammals. Indeed it has been shaped during the evolutionary history of each mammalian species to match the unique developmental needs of its infants. Milk contains numerous compounds that promote infant growth, development, and immune function. Many humans are unusual among mammals in that they consume the milk of different species and consume it well beyond the age at which they would be weaned from nursing. Thus milk has some qualities not shared by other foods, and, as a result, possibly some special physiological effects on humans who consume it.

This "specialness" of milk is manifest in a variety of ways. In the United States and many other countries, milk is featured in food-based dietary guidelines such as the Food Pyramid, where it has its own "food group." As Figure 1.1 shows, the other groups are just that: multiple foods fall within the "fruits," "vegetables," or "meat and beans" groups. Milk and the products made from it are singled out for daily consumption, with two to three servings as the daily recommendation. Milk is also privileged in U.S. government feeding programs such as public school breakfast and lunch programs, and milk is reimbursed by the U.S. Department of Agriculture (USDA) when served in preschools or private schools. It is also part of the Special Supplemental Nutritional Assistance Program for Women, Infants, and Children (WIC), which was initially authorized

Figure 1.1 **The U.S. food pyramid** (Source: www.mypyramid.gov).

in 1974 to subsidize nutrient-rich foods for pregnant or breastfeeding women and infants and children up to 5 years of age. Fluid milk and cheese are featured among the foods that are supported. Across the globe school milk programs are particularly common, and milk is often included in international food donation programs. Thus there is a sense that milk is an especially valuable food, even "essential" to the diet, particularly of children, and that the absence of milk in the diet is cause for concern.

However, another way in which milk is "special" has to do with the sugar in it. Lactose is a sugar found only in milk (hence the term "sweet milk" for fresh milk, as opposed to various fermented "sour" forms of milk), and there is well-documented genetic variation among human populations with respect to the ability to digest this sugar throughout life, with only a minority of humans having this capability. Virtually all humans produce the enzyme necessary to digest this sugar when they are young, but most stop producing it sometime during childhood such that lactose ingestion during adulthood results in a variety of gastrointestinal complaints, which are collectively termed *lactose intolerance*. Historically there has been pronounced cultural variation in the valuation of milk, with some groups idealizing this food and others emphatically disdaining it. These divergent views of milk have tended to map on to variation in milk digestion, with those populations with the genetic trait allowing lactose digestion throughout life being those who strongly favored milk. This range of views about milk has been undermined by efforts to globalize milk production and consumption. More and more countries have jumped on the milk bandwagon despite quite distinct subsistence, culinary, and evolutionary histories.

My goal in this book is to provide insight into this food that can be considered "special" in several respects, and that is currently widely, although by no means universally, believed to be essential to human diets. First and foremost it is important to have a good understanding of what milk is as a food. In later chapters I will take up the question of what kinds of biological effects milk has on humans, focusing on variation of milk digestion and whether milk has growth-promoting effects on children. I will also touch on some of the current controversies surrounding milk consumption among adults (e.g., milk's link to bone density and weight regulation). Framing the discussion of the biological and epidemiological evidence used to evaluate these linkages will be the cultural context that shapes how and why these questions are asked, how the results of this research resonate with or shape our beliefs about milk as a food, and how they are used in milk advertising and the formulation of dietary policies in different countries.

This particular food commodity is of interest in its own right as both a biological and a cultural product, but as a food it can also provide a lens that reveals underlying socio-cultural processes. Milk can be used to illustrate beliefs about what constitutes a "normal" diet or a "healthy" body. Insofar as these beliefs inform interpretations of other diets or bodies and judge them to be inferior, this study of milk yields insights into what I call *bio-ethnocentrism*. This term refers to the interpretation of other people's bodies and behavior only in relation to those of one's own body and culture, generally with the view that one's own is "better" than the other, or that one's own is "normal" and others are deviant or somehow "abnormal," or "pathological."

Milk also serves as a particularly illustrative example of how political and economic processes inform dietary policies and nutrition education. Since their original development, milk has been part of dietary recommendations in the U.S., and milk's status as a "good" or even "essential" food has been largely uncontested. This is in part due to a strong and active dairy lobby that ensures that its product is featured prominently in government-sanctioned recommendations (Nestle 2002). While it may seem that milk would be a poor candidate for global trade, given its tendency toward rapid spoilage, technological innovations and the drive to expand markets for this commodity (and the technology that supports it) have resulted in an unprecedented movement of milk around the world. Thus milk can be used as a study in the globalization of food and Western culture.

A Biocultural Perspective on Milk

To fully consider these questions we need a biocultural approach, one that appreciates milk as both a biological substance with specific physiological effects on the body, and as a food that is part of complex global and local agricultural economies and that has an array of cultural meanings across and within societies. Biocultural approaches draw on the holistic tradition of anthropology, a discipline that seeks to understand the "whole" of humanity. Anthropology includes the study of human evolution, human behavior in the past (archaeology), contemporary cultural variation, and language and linguistic diversity. Each of these aspects of anthropology can contribute something to our understanding of milk, as our experiences with milk are shaped by cultural, political, economic, and evolutionary processes and are expressed through verbal and visual communication.

Of the many perspectives that anthropology can offer any aspect of human behavior, in this book I will generally use approaches from cultural and biological anthropology. Biological anthropologists seek to understand how humans are biologically similar to or different from other organisms, and why humans have their particular biological characteristics (e.g., a large brain, or an anatomy built for bipedalism). To do so they use the principle that unites all of the life sciences – evolution. Evolutionary theory, which dates back to Charles Darwin in the mid-19th century, has provided us with remarkable insights into the biology of all living (and extinct) species, including our own, *Homo sapiens*. It provides a way to understand variation between and within species, both over time and in the present. One of Darwin's main contributions was to demonstrate that the process of natural selection could account for some of this variation. Natural selection simply refers to the fact that individuals vary, variation is heritable (i.e., passed from parents to offspring via their DNA), and that individuals with characteristics that help them survive and/or reproduce in the context of the environment in which they live (which contains predators, food sources, and other living and physical features) will have more offspring than others without those traits. Since these traits are inherited, more individuals in the next generation will have these traits, and over time they should become typical of the population or whole species. Characteristics that enhance survival and reproduction are called *adaptations* or adaptive traits. We expect organisms to have adaptations for getting food, escaping predators, staying warm/keeping cool, coping with pathogens, and so on. In the case of milk, the question is how the introduction of this new food source from domesticated mammals resulted in natural selection favoring genes that would allow for its consumption throughout life.

Life history theory, a branch of evolutionary theory that seeks to understand the allocations of time and energy that organisms make to different parts of their lifecycles (Stearns 1992), is also relevant to the discussion of milk. Life history theory starts from the supposition that organisms have finite time and energy, and they should allocate these resources in a way that is adaptive, i.e., that increases their likelihood of survival and enhances their reproductive success. Since these resources are limited, organisms face two major trade-offs: (1) Should they take a long time to grow to a large body size, and, once they have amassed bodily resources, then reproduce? Or, should they grow quickly and achieve a smaller adult size, but reproduce earlier? (2) Should they have a few offspring and "parent" them or should they have many offspring but invest relatively few resources in any of them? These trade-offs ultimately shape the life history, which is the full description of the various biological states of an individual from conception until death. Organisms' life histories are in part a product of their genetic heritage, but they are also altered by the environment in which they live. Diet is one aspect of that environment that is likely to shape life history.

Relative to other animals, mammals have few offspring and the mother produces milk to nourish them and enhance their survival. Humans take this a step further by provisioning their young after they are weaned from mothers' milk, and among some human populations, the milk of domesticated mammals is fed to children. This raises two interesting life history issues. The first is that milk of any mammal matches its own life history needs; some species have young that are very small at birth, but who grow rapidly and mature early; others produce relatively large infants who grow and mature slowly. Compared to human infants, calves grow very quickly and to a larger size by the time they are weaned from milk. Thus we can ask whether consumption of cows' milk during human childhood alters growth and development in some manner. Second, cows' milk is drunk not only during childhood, but some individuals consume it throughout their life. What are the biological effects of consuming a growth-promoting food during adulthood or senescence, well past the period of growth and development? Since milk is only part of a complex diet, it is difficult to isolate any effects which milk may have on growth or adult biology, but from an evolutionary perspective these are questions that are well worth asking.

Human life histories also unfold within the parameters of a social and cultural environment which, among other things, affects individuals' access to food. In market-based economies, acquiring sufficient, safe, or healthful food is in part a matter of having enough money and also having stores that offer such food close by or easily accessible by car or public transportation. These factors generate a structural framework within which people make decisions about what they purchase, and these decisions are strongly influenced by ideas about particular foods. In the U.S., milk is likely to be found in just about any store that has even a very small selection of food, including convenience stores, corner bodegas, gas stations, large general retailers (e.g., Walmart), and of course grocery stores.

Within these parameters, ideas about foods will influence the likelihood that people will spend their money on them. These include ideas about the biological effects of a given food (e.g., Will it taste good, will it satisfy my hunger, is it good for me, is it good for my children, will it make me sick?). Of course what defines "good" is highly subjective and depends on what currency "goodness" is being evaluated in; it may be nutrients or caloric content, or specific food components such as flavoring, additives, organic or "natural" ingredients, or past experience with the food (e.g., Did I get sick after consuming it last time?). Does it look or smell fresh, and what does "freshness" indicate? (See Freidberg (2009) for a discussion of the idea of "freshness," which is closely associated with milk in the U.S.A.) These are just a few of the culturally informed ideas about food that shape its purchase and ultimately its consumption.

What ideas do people have about milk? What is "good" about this food, [2] and does that vary by age or gender? How does milk consumption affect body size, and how do those ideas mesh with ideals for body size and shape? What authoritative messages about milk are out there, and who are the authorities? How is milk advertised, and what claims are made about it? What other cultural associations does milk have (e.g., biblical references, its "pure" white color, milk as a manifestation of maternal nurturing)? Milk is an aspect of mammalian biology, and its production, promotion, and consumption are all shaped and transformed by human cultural processes. Ultimately these have some impact on the biology of those who drink or avoid drinking milk.

Hence the utility of a combined *biocultural* approach for our study of milk. I see this approach as one in which biological data and evolutionary theory are the focus and framework respectively, but with a clear recognition that human biology is influenced by social and cultural processes that are historically configured. The study of food is particularly amenable to a biocultural lens, as food is clearly a biological substance, transformed through various technologies, and existing in a fluid cultural matrix that predisposes us to consume it (or not). After we put it in our mouths, it is subject to further digestive transformations, and ultimately becomes part of our physiology, with either very temporary or long-lasting effects on our biological functioning. Our biological experiences with food can then feed back into cultural ideals, policies, and economic supports for the foods we desire, or proclaim as "good for us." Although the expression has become somewhat trite, we truly are what we eat, insofar as food becomes our biology and a means by which we embody and express our cultural identity.

Mammals and Milk

How and why do mammalian mothers produce milk for their young? This large taxon of animals is the only group that produces milk to feed their offspring; indeed mammals are taxonomically distinguished from other classes of animals (such as birds or reptiles) by this particular feature. The word "mammal" references the presence of mammary glands (milk-producing glands). Some other animals do feed their young—think of robins bringing worms to the nest to feed their chicks. Honey-bees gather sucrose-rich nectar from flowers and, through repeated cycles of digestion (which break down the sucrose into the simple sugars fructose and glucose) and regurgitation, produce honey. They also bring back protein-rich pollen, which together with honey provides sustenance for the hive. But female mammals are the only organisms that can take their usual diet (whether it be grass, fruits, tubers, nuts, or meat) and transform it into another food entirely, which they can secrete directly to their young.

The process of milk production is complex and takes place in mammary epithelial cells, which are specialized cells in the mammary glands. During pregnancy a rich network of capillaries is formed around these cells to efficiently deliver nutrients and other molecules used to make milk. Various structures within these cells synthesize proteins, lipids (fats), and lactose (a unique milk sugar), which are the macronutrients in milk. Amino acids and glucose are taken up from the bloodstream to make protein and lactose respectively, while some fat is made from circulating fatty acids and other forms of fat are synthesized locally. Other milk components such as antibodies, immune cells, vitamins, minerals, or hormones are taken up from maternal blood circulation, manufactured, or repackaged within the mammary cells.

Making milk is costly to mammalian mothers. Some of their own food intake and stores of energy will be diverted into milk production to feed their young. Why should they do this? From an evolutionary perspective there must be some advantage to this parenting strategy. Providing a rich food source to rapidly growing offspring should increase their survival. In most other animals the vast majority of offspring do not live to reproductive age, but survival rates in mammals are much higher. Because there is lower juvenile mortality and nursing is a time- and energy-expensive behavior, mammals also tend to have relatively few offspring (i.e., they use a "quality" instead of a "quantity" reproductive strategy). Thus milk production is part of an overall mammalian reproductive pattern of increased investment in a smaller number of offspring who have greater chances of surviving to reproduce themselves.

Milk Composition

Just as all mammalian species are different (consider bats, tigers, or whales), so are their milks. Each has a species-specific pattern of growth and development that requires milk with a unique array of nutrients, growth, and immune factors —the baby manatee in Figure 1.2(b) has very different needs from the baby mice in Figure 1.2(a). In addition, the species' environment and feeding ecology will also impact upon milk composition. For example, chimpanzee infants cling to their mothers and nurse "on demand" for several years, while the pups of some seal species are left for several days while their mothers go to sea to forage, and the total duration of nursing may be only a few weeks. Marine mammals living in cold climates need milk that is very high in energy—their milks tend to be very high in fat (many seal milks are over 50% fat, while cows' milk is 4% fat: Oftedal and Iverson 1995). Because infant needs change over the course of development, milk composition changes over time as well. Colostrum, the "milk" produced immediately after birth, is extremely rich in immunoglobulins to protect the infant from infection contracted during birth and the challenges of life outside the womb. Later the protein content diminishes and, as the infant moves toward weaning, nursing frequency diminishes and the volume of milk produced also declines until it ceases altogether when the offspring is fully weaned.

Since this book is about humans drinking cows' milk and the potential biological effects of drinking a milk suited to the growth and development of a very different species, we will focus on this mammal's milk in comparison with human milk. It is worth noting that while the basic components are the same, the milk of contemporary domesticated cattle (mostly *Bos taurus*) may be different from that of their wild ancestors, since cows have been bred for milk production and their diet is often quite different from that of wild cows, as they are

Figure 1.2 **(a) Mouse mother nursing her pups** (Source: *The New Student's Reference Work* (1914)). **(b) Manatee calf nursing** (Source: U.S. Fish and Wildlife Service, Division of Public Affairs).

often fed corn instead of grass. Table 1.1 contains the nutrient composition of human and cows' milk. Cows' milk contains much more protein, minerals (except iron), and some B vitamins, and less sugar, Vitamin C, and Vitamin A. Although you often hear about Vitamin D in association with milk, neither species' milk contains much of this vitamin; commercially available milk is often fortified with Vitamin D, because this vitamin enhances calcium absorption and cows' milk (although not human milk) is very rich in this mineral. Unlike humans, cows can make thiamin in their rumen (actually bacteria living there make this vitamin) and they can also synthesize Vitamin C, so they do not need these vitamins in their diet. The type of fat in the milk of these two species likewise varies. Cows produce more short-chain fatty acids and they also hydrogenate fat in the rumen, producing saturated fat. Human milk has more longer

Table 1.1 **Comparison of the nutrient composition of human and cows milk (per 100 ml)**

Nutrient	Human	Cow
Water (percent)	88	88
Energy (kcal)	70	61
Protein (g)	1.0	3.3
Fat (g)	4.4	3.3
Carbohydrate (g)	6.9	4.7
Minerals		
Calcium (mg)	32	119
Iron (mg)	0.03	0.05
Magnesium (mg)	3	13
Phosphorous (mg)	14	93
Potassium	51	152
Sodium (mg)	17	49
Zinc (mg)	0.17	0.38
Vitamins		
Ascorbic Acid (mg)	5	0.94
Thiamin (mg)	0.014	0.038
Riboflavin (mg)	0.036	0.162
Niacin (mg)	0.177	0.084
Pantothenic Acid (mg)	0.223	0.314
Vitamin B6 (mg)	0.011	0.042
Folate (mcg)	5	5
Vitamin B12 (mcg)	0.045	0.357
Vitamin A (IU)	241	126
Vitamin D	–	–

Source: Adapted from data in Patton (2004).

chain fatty acids, which may be important in the evolution and postnatal development of the large brain (Milligan and Bazinet 2008).

The nutrients in cows' milk are just one set of milk components, and it is wise to keep in mind that a focus on nutrients limits our understanding of any given food. Michael Pollan has termed this limited view "nutritionism." As he notes, "the widely shared but unexamined assumption is that the key to understanding food is indeed the nutrient. Put another way: Foods are essentially the sum of their nutrient parts" (Pollan 2008: 28). A focus on nutrients obscures the fact that there are many other naturally occurring components of foods, such as secondary compounds (chemicals found in plants that are used as defense mechanisms) or hormones that are likely to have effects on our health and well-being. Moreover, depending on the context, nutrients themselves may be perceived as "good" or "bad." Saturated fat, for example, has been vilified in the U.S.A. as a contributor to coronary heart disease and other chronic health problems.

From the perspective of nutritionism, the "goodness" of a food such as milk is attributable to its high density of "good" nutrients: calcium and protein. Milk is also a source of saturated fat, but technological transformations can

Figure 1.3 **A "nutritionist" view of milk** (Source: http://www.floridamilk.com).

customize its fat content to maximize milk's "goodness" in any given context. Figure 1.3 illustrates this with a "nutritionist" image of milk as the sum of its nutrients (although curiously milk's sugar lactose is missing, and fat is labeled "CLA [conjugated linoleic acid] and sphingolipids").

In addition to milk's nutrients there is a variety of immune factors and hormones, including steroids, that act on the adrenal glands and reproductive systems, and those with more restricted activity in the gut, brain, and thyroid. There are also growth factors that stimulate the growth and maturation of a variety of organs. These hormones and growth factors are common to both human and cows' milk, but it is difficult to directly compare amounts in these two species milks because their concentrations vary by stage of lactation and biological condition of the mother. One growth factor that will be discussed in some detail in this book for its potential effects on child growth is insulin-like growth factor I (IGF-I), which is a small peptide in the overall protein fraction of milk. IGF-I is present in both cows' and human milk and the concentration does not vary substantially. IGF-I acts in a wide range of tissues and stimulates growth and differentiation among cells, including bone, and may be involved in the relationship between milk consumption and growth.

The differences between human and cows' milk are more of quantity than of quality, with the most marked difference being the much higher protein and calcium content of cows' milk and some differences in fatty acid types. These reflect the markedly different biology and development of humans and cows. While the overall caloric value of milk does not differ significantly, cows' milk contains about three times as much protein and four times as much calcium, but less lactose and fat than human milk (Prentice 1996). These components allow for a rapid growth rate and the development of a large skeleton (Figure 1.4). Calves grow rapidly in skeletal size and body weight, gaining 0.7–0.8 kg per day throughout lactation in the first year of life (Marlowe and Gaines 1958; Reynolds et al. 1978). In contrast, breastfed human infants gain about 0.02 kg per day in the first year of life (World Health Organization 2007). Of course calves also consume a much larger quantity of milk while nursing as well.

Outline of the Book

In this book I illustrate how one food, milk, has been shaped by cultural, political, economic, and also evolutionary processes and how these have played out historically and cross-culturally. This includes consideration of the historical trends in production, consumption, and marketing, as well as prehistoric trends in animal domestication and the evolution of adult lactose digestion in some human populations. The primary example will be the U.S.A. and to some extent Europe, but we will also look at milk consumption in pastoralist groups in East Africa, and milk in India and China. India is the largest milk producer in the world and privileges milk in a variety of social domains, while China is the

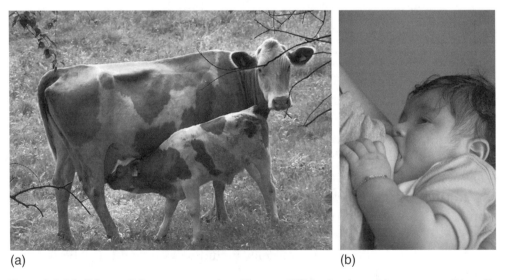

(a)   (b)

Figure 1.4 **(a) Calves of *Bos taurus* nursing** (Source: Wikinoby, http://commons.wikimedia. org/wiki/File:Cow_and_calf.jpg). **(b) Human infant nursing** (Source: Ken Hammond, United States Department of Agriculture.).

current target of global efforts to increase production and consumption despite a history that did not include—and often reviled—cows' milk.

In Chapter 2 population variation in adult lactase production will be covered to better understand why diversity in milk digestion is an important aspect of human biological variation. This is related to the process of animal domestication, and I will review the archaeological data on that process and how/why milking came to be an important part of subsistence in some societies. Population variation in milk digestion is also discussed in relation to contemporary dietary policies that emphasize the importance of milk in everyone's diet. I will review some of the controversies surrounding the nature and importance of this variation in relation to recommendations for milk in the United States and how various interest groups interpret this variation depending on their goals.

In Chapter 3 I will present a brief history of milk consumption in Europe and the United States, to get a better sense of the historical role of milk in the diet, and showing how fresh milk consumption is a relatively recent phenomenon. I will then focus on the United States during the 19th and early 20th centuries to understand how milk came to be constructed as an essential part of the American diet, especially for children. In the 19th century, milk was more likely to be a cause of child illness than superior health, yet by the early 20th century it was already heralded as the ideal food, enhancing the growth of children of all ages. However, in the post-World War II period milk consumption began to taper off, and I ask whether the trend of drinking fresh milk may be considered a brief historical interlude.

A strong link between milk consumption and child growth, particularly in height, has been widely assumed and proclaimed, yet relatively little evidence exists for an association. Does milk consumption enhance child growth? What role does calcium play in this process? I will review the scientific literature on the relationship between milk and growth in Chapter 4 and touch upon the relationship between milk and bone density and weight. I will also look at some of the marketing claims that have been made about milk and body size, and consider these in relation to the socio-cultural context in which they are made.

In Chapter 5, current trends in the globalization of milk will be presented and considered in light of population variation in milk digestion, and how various countries have used assumptions about milk's contributions to better health, nutrition, and growth to further nationalist political and economic agendas. International patterns of milk production and consumption are experiencing a major transition, with declining consumption in traditional milk-drinking countries and skyrocketing intake in those with culinary traditions that historically excluded milk. Why have so many populations who have not historically consumed milk suddenly embraced this food, while those with long histories of milk consumption are abandoning it? What meanings does milk have in these contexts? The book will conclude in Chapter 6 with a critical look back at "nature's perfect food" and the questions that remain about milk's contributions to human health and well-being.

One last note. As some of you may know already, there is ongoing debate about the goodness of milk, and it tends to be highly polarized, with some groups (such as the People for the Ethical Treatment of Animals [PETA]) arguing that milk contributes to a wide range of health problems, and others (such as the National Dairy Council) arguing that numerous contemporary health problems may be attributed to the fact that Americans don't drink enough milk. At times the debate has been ugly, with hyperbole and name calling on both sides, and it remains highly volatile. Then there is the anti-pasteurization movement, which proclaims the virtues of unpasteurized ("raw") milk, and runs up against the Food and Drug Administration's assertion that unpasteurized milk is inherently unsafe. In addition, there are the myriad other "milks" out there (soy, rice, almond, hemp) vying for a share of the milk market and marketed as plant-based alternatives to mammalian milk.[3] Finally, while the dominant ideology overwhelmingly favors milk in the U.S., consumption practices are not always in sync, as consumption has been declining for the past 50-plus years.

Although I am often asked about my feelings about milk, I do not take a strong "pro" or "anti" milk view. I want to better understand both the science underlying claims for milk's value as a food in the human diet, and the social, cultural, political, and economic forces at work to support or undermine certain views about milk. Milk is a fascinating food; it reveals as much about sociocultural processes as it does about the evolutionary processes that gave rise to milk production and consumption in the first place.

2

POPULATION VARIATION IN MILK DIGESTION AND DIETARY POLICY

Milk contains a unique sugar, lactose, which turns out to be more interesting than it might first appear. The ability to digest this sugar varies across the lifecycle and among populations, and it turns out that drinking milk presents a challenge to many of the world's populations. In this chapter I describe the biology of lactose digestion and population variation in this trait, and then turn to how this diversity intersects with dietary policies in the U.S. that encourage milk consumption. The linguistic terms employed in discussions of this variation vis-à-vis milk promotion provide insight into how different institutions involved in designing or implementing dietary recommendations view this important aspect of human biological diversity.

Lactose and Lactase

As we saw in Chapter 1, most, though not all mammalian milk, including that from humans and cows, contains lactose. As shown in Figure 2.1, lactose is a disaccharide—a double sugar made up of two simple sugars: glucose and galactose. As such, lactose cannot be absorbed into the small intestine directly. Instead it must be cleaved into the two single sugars, which can then be absorbed, enter the body's circulatory system, and used for energy. This process requires a specialized enzyme called lactase (more technically, lactase-phlorizin hydrolase [LPH; also abbreviated as LCT]), which is produced in the pancreas and secreted from there into the small intestine. Among mammals whose milk contains lactose, infants produce lactase in order to digest the lactose they are ingesting while nursing. However, lactase production diminishes over time and eventually stops altogether, usually around the time of weaning, as shown in Figure 2.2. Importantly, mammals living in the wild never consume milk again after they are weaned; since the sole function of lactase is to cleave lactose, it would be wasteful of scarce energy and nutrient resources to continue to produce a useless enzyme.

The gene for the lactase enzyme is on chromosome 2. The region upstream from the lactase gene has been identified as a site of variation in adult lactase activity (Enattah et al. 2002; Ingram, et al. 2007, 2009; Tishkoff et al. 2007).

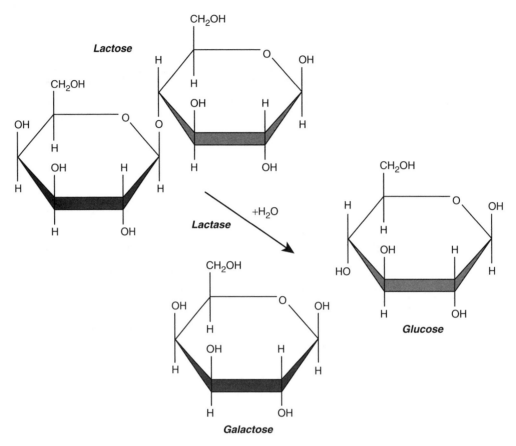

Figure 2.1 **The chemical structure of lactose and action of lactase.**

Some human populations have high frequencies of mutations in this region that cause the lactase gene to remain active throughout life. In particular, some sub-Saharan African, European, and Middle Eastern populations have shown high frequencies of these mutations, which vary across populations (see Ingram, et al. (2009) for a full discussion of the genetic variants associated with

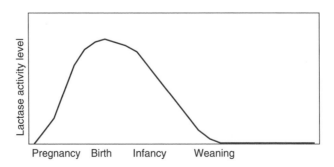

Figure 2.2 **Life history pattern of lactase production for most mammals.**

lactase persistence).[1] The mutations in the lactase gene that result in lactase remaining "on" throughout life appear to be dominant to those that result in lactase being shut off around the time of weaning, and thus only one copy of the mutated gene is needed to continue to produce lactase into adulthood.[2]

The signals that control changes in lactase activity are not well understood. Up until the latter part of the 20th century, researchers debated whether lactase activity could be induced by a diet containing lactose; that is, they wondered whether lactase activity could be turned on and off by the presence or absence of milk in the diet. Although a few studies suggested that it could be, most concluded that the decline of lactase production occurred independent of the presence of lactose in the small intestine (Sahi 1994b), and that drinking milk would not cause lactase to be continually produced. The mechanisms underlying the down-regulation of lactase activity are the subject of ongoing scientific investigation.

All groups that have high frequencies of the mutations for lifelong lactase activity share a history of dairy animal domestication and reliance on milk from these animals. Dairy animals were domesticated in the Old World around 8,000 years ago (see Chapter 3 for more discussion of the origins of dairying), and the mutation for ongoing lactase production appears to have spread after that time (Burger et al. 2007). It is easy to hypothesize that mutations which allowed people to digest milk throughout their lives would have been advantageous in Old World dairying populations, as they would have been able to utilize a nutrient-rich food. Individuals with the mutation would have been better nourished and presumably healthier, and ultimately produced more offspring who also had the mutation (Ingram et al. 2009; Simoons 1978, 2001; Tishkoff et al. 2007). Natural selection would have caused this mutation to spread rapidly throughout these populations. Among populations in the New World, East Asia, Oceania, or other areas of Africa who did not domesticate milk-producing animals or drink their milk, ongoing lactase production would have offered no advantage, since there was no milk in the diet after weaning. As a result, most humans maintain the ancestral mammalian pattern in which lactase activity is turned off during the juvenile period.

Milk is certainly a nutrient-rich food, but it is also possible to make use of these nutrients by employing simple processing techniques that reduce the lactose in milk, thereby circumventing the need for continual lactase production. Allowing fresh milk to ferment in warm climates causes lactophilic bacteria (such as *Lactobacillus*) to break down the lactose, producing yogurt. Depending on how long this process goes on and the quantity of bacteria present, most of the lactose will be removed, leaving the tangy taste of lactic acid characteristic of yogurt. This is the most common means by which milk is processed around the world (Mendelson 2008). Alternatively, adding a strong acid to heated milk will cause it to curdle, as the milk solids separate from the liquid whey portion. This basic process produces cheese. Lactose is dissolved in the liquid whey that is drained off; hence cheese has little lactose in it. In general, the softer the cheese the more whey is retained in it, and the higher the lactose content. A very dry cheese such as parmesan contains virtually no lactose; mozzarella contains much more.

Given that these are very basic processing techniques, and that they result in dairy products with little lactose but many nutrients left in them, why did the mutation for ongoing lactase production spread? It is worth noting that culinary behaviors can spread very rapidly through observation, teaching, and diffusion, while genetic change is relatively slow, with advantageous genes spreading through the population over many, many generations. William Durham (1991) observed that populations with high frequencies of adult lactase activity not only keep dairy animals, but they also drink *fresh* milk. Fresh milk consumption clusters in two areas of the world: Northern Europe and the areas of sub-Saharan Africa inhabited by dairying populations.[3] Durham proposed that the low levels of UV light found in the high latitudes of Northern Europe would have selected for ongoing lactase activity among dairying populations living there. This hypothesis was based on the finding of Gebhard Flatz and Hans Rotthauwe (1973) that the presence of lactose in the small intestine enhanced calcium absorption. Under high UV light conditions Vitamin D, which is synthesized in skin cells in the presence of UV light, facilitates calcium uptake. When exposure to UV light is reduced and Vitamin D synthesis is likewise diminished, lactose, which is found in fresh milk, but in only trace amounts in other dairy products, can increase calcium absorption. Although this is an intriguing hypothesis, geneticists have found no evidence for an effect of latitude (as a marker for availability of Vitamin D synthesis) on the spread of mutations for adult lactose digestion in Europe (Beja-Pereira et al. 2003).

While this mechanism may be in play among Northern Europeans, it has little relevance for understanding why some populations at lower latitudes continue to produce lactase throughout life. There may have been other nutritional advantages to adult milk consumption for sub-Saharan herders. These may have included the use of fluid milk for hydration; sources of water are scarce in the arid areas where animals are pastured, and place an important constraint on human use of these landscapes. In addition, carbohydrates are limited in the diets of pastoralist peoples. While animal products contain myriad micronutrients, protein, and some fat, they do not contain carbohydrates (sugars), which are the preferred source of energy for human cells. Indeed an important difference between animals and plants is that animals store energy almost exclusively as fat (there are very limited amounts of a stored carbohydrate called glycogen, which is used for immediate muscle needs), while plants store energy primarily as carbohydrates; they do need fat for use as a weather-resistant form of energy for seedlings. Thus for populations relying heavily on animal products, the only way to access carbohydrates is to drink the milk of domesticated mammals.

It is important to recognize that continual lactase production is the exception, rather than the rule, for our species; that is, most human populations continued the basic mammalian pattern of down-regulation of lactase during childhood. This is demonstrated in Table 2.1, which shows percentages of individuals with lactase activity in adulthood from a sample of populations from different parts of the world. While frequencies of adult lactase activity are somewhat continuously distributed across populations, two patterns are easily distinguished: (1) populations in which high frequencies of adults continue to produce high levels of lactase in adulthood; and (2) populations in which lactase production declines to low levels by adulthood (Sahi 1994a). The latter are much more numerous. Populations with intermediate frequencies tend to be those in which there has been substantial intermarriage among individuals from populations with high and low frequencies. Note also that in countries that have been colonized by peoples of European descent (the United States, Australia, New Zealand, and to varying degrees countries in Latin America), rates of adult lactase production tend to be high or intermediate. Populations also vary in the age at which lactase activity declines, from one to two years to 20 years (Sahi 1994a). Neither the mechanism underlying this age variation nor its significance is well understood.

Early Work on the Clinical Significance of Population Variation in Milk Digestion

An understanding that populations varied with respect to lactase activity and that most humans retained the basic mammalian pattern in which lactase production stopped around the time of weaning was not fully appreciated up through

Table 2.1 **Prevalence of lactase persistence in various human populations**

Population	Percentage
Northern Europeans	85–97
European-Americans	78–94
Central Europeans	77–91
South Asians	
Northern South Asia	70–80
Southern South Asia	30–40
Hispanics	20–50
African-Americans	20–40
Native Americans	0–20
East and Southeast Asians	0–5

Source: Adapted from data in Sahi (1994).

the 1960s. Clinicians and nutritionists in the United States worked under the assumption that everyone could and should drink milk. Anyone who experienced negative symptoms such as nausea, diarrhea, gas, bloating, or general gastrointestinal discomfort after drinking milk was considered to have some underlying pathology. The "normal" human physiology included the capacity to digest milk throughout life. This view made sense to people who were largely of European descent—after all, they had no trouble digesting milk. This is an example of what I referred to in Chapter 1 as "bio-ethnocentrism," the belief in the superiority of a biology and set of cultural practices, usually one's own. In this case, the idea that everyone should drink milk and have the digestive biology that allows for milk digestion without unpleasant symptoms is an example of bio-ethnocentrism.

The formative studies demonstrating population variation in milk digestion came from researchers in Baltimore, MD, who found that 75% of healthy adult African-American men but only 10% of European-American adult men (both groups comprised volunteers from a correctional facility) had "lactase deficiency" (Bayless and Rosensweig 1966). Furthermore, over 90% of the African-American group reported the symptoms of lactose intolerance, while only 10% of the European-Americans did so. Individuals with low levels of lactase who reported symptoms of lactose intolerance had no other signs of intestinal pathology. Studies also confirmed that lactase activity in adults could not be increased by providing lactose (Rosenzweig 1973). Subsequent studies revealed a similar result: most humans have lactase turned off during childhood (Bayless and Rosensweig 1966, 1967).

These authors also speculated on the role of milk in the history of African populations from which African-Americans are largely descended, using the terminology of the time: "Milk drinking after weaning is very unusual in the areas of central west Africa, whence the American Negro came" (Bayless and Rosensweig 1966). They were also quite concerned that school lunch programs which provided milk to children were potentially posing a problem for African-American children, many of whom they observed throwing away their milk or drinking little of it. They noted:

> one should recognize that lactose malabsorption may be an additional factor contributing to milk rejection on the part of Negro children. One may not, therefore, assume that a half-pint carton of milk, if made available and placed on a lunch tray, will be consumed. Current results suggest the need to reconsider the rationale of attempts to reinforce the nutritional status of Negroes through heavy emphasis on milk consumption and milk distribution.
>
> (Paige et al. 1972: 1488)

Thus recognition that ethnic minorities might be adversely affected by dietary policies that promoted milk consumption in the U.S. due to genetic differences in lactose digestion has existed since the early 1970s, and I will consider its current relevance in subsequent sections.

Terminology

A number of different linguistic terms are employed to describe the biological phenomenon of lactase activity in adults, and their usage provides insights into how various authors or institutions perceive biological difference and its significance for milk consumption. That is, these terms allow us to see if bio-ethnocentric views persist in current discussions of this biological and dietary variation. In vernacular usage, the term lactose intolerance is most commonly used. This refers specifically to the subjective experience of gastrointestinal symptoms after lactose consumption (e.g., bloating, diarrhea, cramps, gas). As it turns out, these symptoms vary considerably across individuals with different levels of lactase activity. It is quite possible to have high lactase activity but to report intolerance, or to have biologically assayed low lactase activity but to report no symptoms of intolerance. Given that "intolerance" is in part a matter of subjective experience, other terms better describe the biological foundation of variation in milk digestion, although there is no consensus on which term is preferable.

The term *adult-type hypolactasia* has gained currency as a way to describe low levels of lactase activity among adults. However, this term is problematic insofar as it suggests that the modal condition for humans is "too little" lactose (i.e., *hypo*lactasia). Sahi, a leading proponent of this terminology, dismisses the possibility of using *hyperlactasia* as its counterpart (Sahi 1994b), but given that this is an unusual condition for our species, individuals with lactase activity in adulthood could properly be described as having *hyperlactasia*. The early work on population variation in lactose digestion described in the previous section routinely used the term *lactase deficiency*. Although those authors argued that this condition was the norm for many populations, they retained this phrase, which implies pathology, as do the descriptors *lactose maldigestion* and *malabsorption*, which are currently widely used (*mal* comes from the Latin for bad or wrong). It is also worth pointing out that terminology including the word *lactose* implies that lactose is part of the diet; lactose maldigestion or malabsorption would never manifest if lactose was not being consumed.

I will use the terms *lactase persistence* and *lactase impersistence* throughout the book, unless otherwise noted. These terms are preferred because they are relatively value-free, implying neither pathology nor having *too much* or *too little* or *high* or *low* lactase activity and they incorporate the fact that the variation is in terms of how long lactase is active (i.e., it persists) rather than if an individual does or does not produce lactase. As we will see later in this chapter, usage of the terms lactase persistence or impersistence is not widespread beyond anthropology, either in the vernacular or in the writings of a variety of professional organizations that are concerned with health or nutrition.

Changes in lactase activity or the digestion of lactose among adults may also stem from nutritional factors (e.g., protein malnutrition), pathologies of the small intestine, and gastrointestinal pathogens. When any of these result in low levels of lactose digestion, secondary hypolactasia, malabsorption, or maldigestion is the diagnostic label. Adult-type hypolactasia that derives exclusively from genetically regulated declines in lactase activity (i.e., lactase impersistence) is considered primary.

Dietary Policy, Population Variation, and Milk Consumption in the U.S.

How do institutions and organizations involved in the construction and translation of dietary policy confront the issue of population variation in milk digestion, especially in a cultural context in which milk is highly valued, both as an important part of the agricultural economy and as a particularly nutritious food? Let us first look at how U.S. dietary policies are constructed, the role of milk in such policies, how the issue of population variation in milk digestion is accommodated, and what terms are used to describe this variation. Official U.S. government dietary policy is articulated in the Dietary Guidelines for Americans (United States Department of Health and Human Services and United States Department of Agriculture 2005). These guidelines are updated every five years through a complex process that includes the United States Department of Agriculture (USDA), Department of Health and Human Services (DHHS), expert panelists, public comment, and industry lobbying. Readers are probably most familiar with the USDA Food Pyramid, which is the pictorial representation of the Dietary Guidelines for Americans (see Figure 1.1).

The USDA and Milk Promotion

It is worth considering the history and purview of the USDA. This entity has a dual mandate: to promote U.S. agricultural interests and to issue food and nutrition guidelines that promote the health of U.S. citizens. That these two missions may be at odds with one another was apparently not considered when the USDA first started issuing dietary advice in the early 20th century. At the time, undernutrition was a considerable problem. However, in the current dietary environment of the hyperabundance of relatively cheap agricultural commodities (e.g., corn, wheat, milk), conflict between these goals is becoming more visible as rates of obesity and its concomitant health problems continue to rise (Nestle 2002). Despite the surfeit of food now available to and consumed by

Figure 2.3 **Donna Shalala in one of the popular "got milk?" advertisements.**

Americans and problems of overconsumption, the USDA must continue to support the consumption of U.S. agricultural commodities. Various agricultural groups such as the dairy industry are influential in crafting dietary guidelines (Nestle 2002), as these recommendations have the potential to dramatically impact the market for agricultural commodities. In what some saw as a blatant conflation of government policy and corporate advertising was the 1998 appearance of then-Secretary of Health and Human Services Donna Shalala in one of the popular "got milk?" advertisements (see Figure 2.3).

The dairy industry makes up about 11% of the U.S. agricultural economy (United States Department of Commerce 2002). In the Dairy Production Stabilization Act of 1983, the U.S. government authorized the USDA to oversee national programs for "dairy product promotion, research, and nutrition education as part of a comprehensive strategy to increase human consumption of milk and dairy products" (United States Department of Commerce 2002: 5). In 1990, the Fluid Milk Promotion Act specifically targeted fluid (fresh) milk. In justifying this Act Congress stated that:

(1) fluid milk products are basic foods and a primary source of required nutrients such as calcium, and otherwise are a valuable part of the

human diet; and (2) fluid milk products must be readily available and marketed efficiently to ensure that the people of the United States receive adequate nourishment; and (3) the dairy industry plays a significant role in the economy of the United States.

(United States Department of Agriculture 2010a)

These programs are run by groups such as the Fluid Milk Board, the National Dairy Council (NDC),[4] and the National Dairy Promotion and Research Board (NDPRB), among other state and regional organizations. Under a check-off system, local dairy farmers pay a mandatory fee per unit of milk produced to support the activities of these groups, the vast majority of which focus on advertising (United States Department of Agriculture 2002).

As we saw in Chapter 1, milk has a dedicated "group" in the USDA Food Pyramid, and two to three servings of milk or an equivalent amount of dairy products are recommended for daily consumption. The primary (though not exclusive) justification for such a recommendation is based on milk products as rich sources of calcium. Moreover, even though some "alternative" non-official food pyramids are designed for minority populations that are accessible through the USDA site, all of them contain dairy products (USDA National Agriculture Library Food and Nutrition Information Center 2010). As Marion Nestle (2002) noted in her careful study of links between government diet and nutrition policy and food industries, several representatives of the dairy industry including the NDC and NDPRB were on the advisory committee charged with developing the 2000 Dietary Guidelines for Americans. Not surprisingly, they opposed any suggestion to include alternatives to dairy foods, especially as sources of calcium (such as fortified soy milk) in the milk group. The inclusion of soy milks may be more acceptable now to the dairy industry as the largest milk processor in the U.S., Dean Foods, has bought several soy milk companies such as White Wave (which makes the brand Silk®). The 2010 update of the Dietary Guidelines is underway, and it will be interesting to see if any of the language about non-dairy milks is different.

Thus in both of its roles the USDA is involved in promoting the consumption of dairy products—especially fresh milk—among all U.S. citizens. However, the agency has been forced to recognize that substantial numbers of ethnic minorities may be lactase impersistent and/or consider themselves lactose intolerant. With the increasing presence of peoples of Asian, Latin American, or African descent in the United States, up to 25% of the adult population may be lactase impersistent. The National Dairy Council recognizes that approximately 100% of all Native Americans, 90% of all Asian Americans, 80% of all African Americans, 53% of all Hispanic Americans, yet only 15% of all "Caucasians" are "lactose maldigesters" (National Dairy Council 2003b).

The 2005 Dietary Guidelines acknowledged the existence of lactose intolerance among U.S. citizens and offers some alternatives:

> If a person wants to consider milk alternatives because of lactose intolerance, the most reliable and easiest ways to derive the health benefits associated with milk and milk product consumption is to choose alternatives within the milk food group, such as yogurt or lactose free milk, or to consume the enzyme lactase prior to the consumption of milk products. For individuals who choose to or must avoid all milk products (e.g., individuals with lactose intolerance, vegans), nondairy calcium containing alternatives may be selected to help meet calcium needs.

These alternatives (listed in an appendix) include various fortified products (cereal, soy milk, orange juice), along with tofu, greens, and fish, among others (United States Department of Health and Human Services and United States Department of Agriculture 2005). However, these are prefaced with a caveat on the mypyramid.gov website: "Calcium-fortified foods and beverages such as soy beverages or orange juice may provide calcium, but may not provide the other nutrients found in milk and milk products."

The USDA also administers a number of federally funded programs that promote milk consumption. Since the 1946 inception of the National School Lunch Act, the government has required that fluid milk be offered as part of meals that are eligible for federal reimbursement. Twenty years later, private institutions devoted to the care and training of children were also made eligible for these federal milk reimbursements. In 1968 an amendment to the Child Nutrition Act was approved that read: "Minimum nutritional requirements shall not be construed to prohibit substitution of foods to accommodate the medical or other special dietary needs of individual students" (USDA Food and Nutrition Service 2009). Presumably, "special dietary needs" covered those with lactose intolerance.

The Special Supplemental Nutritional Assistance Program for Women, Infants and Children (WIC) is also under USDA authority. The program began in 1974 to provide subsidies for specific nutrient-rich foods for pregnant or breastfeeding women and infants and children up to 5 years of age. Fluid milk and cheese are featured among the foods that are allowed (others are infant formula, cereal, eggs, dried beans, peanut butter, tunafish, and carrots). The majority (over 60%) of WIC recipients (which numbered almost nine million in 2008) are minorities, the largest percentage of which are Hispanics, African-Americans, Asians or Pacific Islanders, and Native Americans, all populations with high frequencies of lactase impersistence (USDA Economic Research Service 2002). In recognition of individuals with "special" dietary needs, WIC allows lactose-reduced or lactose-free milk, or the substitution of more cheese for milk in its food packages. An interim rule implemented in 2007 allows fortified soy-based beverages and tofu, as well as fruits and vegetables, juice, whole-wheat bread, and canned fish. These changes were put into place to align with the 2005 Dietary Guidelines and to accommodate cultural food preferences of WIC participants (USDA Food and Nutrition Service 2010).

Population Variation in Lactase Production: Dairy Industry Perspectives

The terminology used by dairy promotion organizations such as the National Dairy Council is indicative of their interest in promoting milk consumption in light of awareness of diversity in lactase production among U.S. citizens. First, a clear distinction is made between the terms *lactose maldigestion* and *lactose intolerance*. As defined earlier, lactose intolerance refers to the "gastro-intestinal symptoms experienced by some individuals who have low levels of lactase, the enzyme necessary to digest lactose" (National Dairy Council 2003b). However, they argue that this condition is relatively rare, and that *lactose maldigestion* is more common. Importantly, individuals with lactose maldigestion are considered to be individuals who may have low levels of lactase but who do not experience gastrointestinal symptoms following consumption of lactose-containing dairy products so long as their physiological capacity to digest lactose is not exceeded. Thus, according to the NDC, such people can—and should—consume milk.

The NDC contends that rates of reported lactose maldigestion are likely to overestimate the number of individuals who actually suffer negative symptoms after milk consumption, in part because these symptoms may mimic those of other gastrointestinal illnesses. Therefore the NDC recommends that individuals who suspect they are lactose intolerant should be tested by a physician using the breath hydrogen test, which measures the amount of hydrogen expelled in a person's breath following the digestion of lactose. The organization is, however, careful to note that those tests may generate false positives (again inflating the rate of "true" intolerance) because the lactose challenge is much greater than that found in a glass of milk. Furthermore, many individuals may claim to be lactose intolerant not because they have physiological symptoms but rather due to negative "culturally-based attitudes towards milk learned at a young age" (National Dairy Council 2003b). Such persons may never acquire a taste for milk if they come from a family that does not make milk a part of their regular diet.

In an effort to encourage milk consumption among lactase-impersistent individuals, materials available through the USDA and dairy promotion agencies recommend "several easy steps to overcome lactose intolerance." The first and most vital step is to see a physician immediately in order to be correctly diagnosed.

If a low level of lactase activity is verified, an individual must *not* conclude that he or she should avoid dairy products but rather find creative ways to include dairy products in their diet. The NDC warns that "avoiding dairy foods can cause inadequate intakes of calcium and many other essential nutrients. A deficiency of calcium increases the risk of developing osteoporosis, hypertension, and possibly some types of cancer," but "fortunately, tolerance to lactose can be improved by adjusting the amounts and types of dairy foods consumed" (National Dairy Council 2003a). These modifications include drinking small amounts of milk with meals to slow the process of absorption or starting with small servings and slowly working up to larger quantities to "build" tolerance to lactose. Other solutions to intolerance include consuming aged hard cheeses, yogurt with active bacterial cultures, lactose-free milk, or taking over-the-counter lactase enzyme tablets or drops prior to the consumption of lactose-containing dairy products. Non-dairy sources of calcium are denigrated as having much less calcium than milk, or it is suggested that the calcium in certain vegetables is much less bio-available than that of milk.[5] Thus the USDA and the National Dairy Council seek to minimize the prevalence of lactose intolerance, and advocate ways for individuals with *lactose maldigestion* to overcome this "problem" and include more dairy in their diet.

The significance of biological variation in lactase persistence is downplayed by the dairy industry. They consider it to be an "overblown" issue, citing evidence that lactase impersistence is not always associated with symptoms of lactose intolerance. Although they acknowledge that a large portion of adults—particularly in minority populations in the United States—are likely to have low levels of lactase, this should not prevent them from consuming milk. Research sponsored by the NDC has been focused on determining just how much milk people who self- or medically diagnose lactose maldigestion can consume. Their published studies show that up to two or even three cups of milk may be consumed by individuals testing positive for lactose maldigestion, as long as these cups are spread throughout the day (Suarez et al. 1997, 1998). Anthropological studies have also noted discordance between lactase status and symptoms of intolerance. For example, Susan Cheer and John Allen (1997) found that Tokelau islanders in New Zealand had high frequencies of lactase impersistence, as diagnosed in breath hydrogen tests, but lactase status was not highly correlated with either consumption of dairy products or perceived symptoms of lactose intolerance.

Dietetics and Medical Perspectives on Variation in Lactose Digestion

Nutritionists and dieticians also translate federal nutrition policy to individuals. Organizations such as the American Dietetic Association (ADA), the professional unit to which registered dietitians belong, and nutrition textbooks tend to mimic the approach of the USDA and NDC. Given that nutritionists and dieticians are those who provide dietary advice to individuals in clinical, public health, food assistance programs, and other settings, their interpretation of lactase impersistence is likely to have very tangible effects. As is clear in the following descriptions, they have tended to adopt the same terminology and perspective of the USDA and National Dairy Council.

A popular nutrition text (Wardlaw and Hampl 2007) discusses the topic in the section "Health Concerns Related to Carbohydrate Intake." The authors use the term "lactose maldigestion," noting that it is a "normal pattern of physiology that often begins to develop after early childhood ... This ... is estimated to be present in about 75% of the world's population, although not all these individual experience symptoms. (When significant symptoms develop after lactose intake, it is then called lactose intolerance)" (Wardlaw and Hampl 2007: 173). They emphasize that most people with lactose maldigestion can tolerate milk with meals or other milk products and that it is unnecessary to restrict intake of these foods, as "these calcium-rich food products are important for maintaining bone health" (Wardlaw and Hampl 2007: 174).

The American Dietetic Association (ADA) provides a document "The lowdown on lactose intolerance" on its website (www.eatright.org). This document is actually authored by the National Dairy Council, indicating a close relationship between the dieticians who provide dietary counseling and guidance and the marketing arm of the U.S. dairy industry.[6] Indeed on their website page "Food and Nutrition Information for Consumers" there are multiple links to the National Dairy Council. The link to the NDC material is new: five years ago the ADA had a fact sheet available (sponsored by MacNeil Nutritionals, the producer of Lactaid® products for people with lactose intolerance) which stated that lactose intolerance was quite common (Wiley 2004).

What about clinicians, who provide the authoritative diagnosis of lactose activity? Among professional medical associations, such as the American College of Gastroenterology (AAG: gastroenterology is the study of the gastrointestinal system; members of this organization deal with gastrointestinal diseases), the American Academy of Family Practitioners (AAFP), and the American Academy of Pediatrics (AAP), we see a discrepancy between acknowledgment that lactase impersistence is the norm and portrayal of lactase impersistence as a "disease" or as the "abnormal" condition. First, these groups all make the crucial distinction between lactose intolerance and lactase impersistence, but tend to focus on lactose intolerance. This makes sense because, from their perspective, lactose intolerance is the relevant clinical condition; individuals experiencing uncomfortable or painful symptoms due to underlying lactase impersistence would be those most likely to seek medical help.

For example, the AAP now uses the language *"lactose malabsorption,"* but in the discussion of prevalence, it refers to *lactase deficiency*:

> Approximately 70% of the world's population has primary lactase *deficiency*. The percentage varies according to ethnicity and is related to the use of dairy products in the diet, resulting in genetic selection of individuals with the ability to digest lactose. In populations with a predominance of dairy foods in the diet, particularly northern European people, as few as 2% of the population has primary lactase deficiency. In contrast, the prevalence of primary lactase deficiency is 50% to 80% in Hispanic people, 60% to 80% in black and Ashkenazi Jewish people, and almost 100% in Asian and American Indian people.
>
> (Heyman and and the Committee on Nutrition 2006,
> emphasis added)[7]

Likewise, lactase impersistence is described by the ACG as: "a *shortage* of the enzyme lactase, which is *normally* produced by the cells that line the small intestine" (American College of Gastroenterology 2009a, emphasis added). Note that these are descriptions of lactase impersistence, not specifically the clinical symptoms associated with lactose intolerance. Thus, while recognizing that most humans are lactase impersistent, medical groups tend to—perhaps not surprisingly—"medicalize" this "condition." Medicalization refers to the construction of a bodily state as pathological, with recognizable symptoms and in need of treatment (Wiley and Allen 2008). To clinicians, lactase impersistence may be the norm, but it is still not considered "normal."

In their educational materials, both the AAFP and ACG recommend that individuals diagnose themselves by eliminating all dairy products from their diet for several weeks to ascertain whether this eases their symptoms (this strategy is not recommended by the National Dairy Council, which urges people to get tested for lactase activity). This is followed by a dairy challenge to see if symptoms reappear. If they do, the "treatment" is simple: avoid dairy products. However, they also suggest that through trial and error individuals should figure out how much of which dairy products they can tolerate without experiencing negative symptoms.

The worry for those who avoid milk products is that they would not meet their calcium needs. Most medical organizations recommend dairy products such as yogurt, cheese, or lactose-reduced milk, especially for children, who are seen as particularly in need of not only the calcium in dairy products but also the protein, Vitamin D, and, in the case of fresh milk, water for hydration (Heyman and and the Committee on Nutrition 2006). Other fortified foods such as orange juice or dark green leafy vegetables, legumes, and fish are recommended, and calcium supplementation is advised for those who "significantly limit their dietary intake of milk products" (American College of Gastroenterology 2009a).

While the AAP expresses concern that children with lactase impersistence obtain sufficient calcium, their statement also outlines the potential problems associated with milk consumption among such children. The nutrients in milk may not be fully absorbed; if diarrhea results, nutrients are lost and there is a risk of dehydration. Furthermore, the AAP and others earlier expressed concern about the use of lactose in medicines such as birth control pills, antacids, and other prescription and over-the-counter drugs (American Academy of Pediatrics 1985). Lactose is used as filler, an anti-caking agent, and makes pills more palatable. While only a very few individuals with severe intolerance are likely to be sensitive to these small amounts, including lactose among the inactive ingredients in medicines (estimated at 20% prescription drugs and 6% of over-the-counter medicines: American Gastroenterology Association 2009b) this indicates a lack of appreciation for population diversity in physiological responses to lactose and the surfeit of lactose available from dairy production.

Anti-milk Groups' Perspectives

While the National Dairy Council tends to downplay the issue of population variation in milk digestion, and organizations such as the AAP express concern about "hidden" lactose or the ill-effects of milk consumption among those with lactose intolerance, there is also vigorous—if not well-coordinated—anti-milk sentiment, indicating that milk's merits are not entirely uncontested within the United States. Two primers for this "movement" with intentionally sensationalist titles are: (1) *Milk: The Deadly Poison* by Robert Cohen (1997) (the self-proclaimed "notmilkman" who also maintains a website at www.notmilk.com) and (2) *Don't Drink Your Milk! The frightening new medical facts about the world's most overrated nutrient* by Frank Oski (1977). In addition, the Physician's Committee for Responsible Medicine (PCRM) and People for the Ethical Treatment of Animals (PETA) are two very active organizations promoting the message that milk is neither an ideal nor necessary food.[8] Both groups cite studies implicating milk consumption as a contributing factor to numerous health problems (from prostate and breast cancer to osteoporosis; see www.pcrm.org or www. milksucks.com). Lactose intolerance is on the list of potential problems associated with milk consumption; as with clinicians, anti-milk groups are more concerned with negative physiological outcomes and less interested in lactase impersistence per se except insofar as it may be used to emphasize that this is the norm for our species.

Unlike the dairy industry, which claims that the prevalence of lactose intolerance is overestimated, Cohen suggests that it is under-diagnosed as a source of gastrointestinal complaints. In addition, instead of following the "simple steps" that the dairy industry outlines for individuals with symptoms of intolerance, the solution is simple: avoid milk products. Since the anti-milk contingent considers there to be sufficient evidence that milk may cause rather than prevent various health problems, this avoidance is not deleterious but is, in fact, beneficial to one's health.

Milk Promotion, Population Variation, and Charges of Racism

The clash between the PCRM and the pro-milk agenda of the NDC came to a head in two articles in the *Journal of the National Medical Association*, a journal directed at the needs of African-American health practitioners. At the heart of the debate was the policy significance of biological variation in adult lactase production. Authors from the PCRM alleged that the *Dietary Guidelines for Americans* are biased against minorities insofar as dairy products are recommended for all Americans (Bertron et al. 1999). They concluded that the *Guidelines*:

> encourage dairy products for daily consumption by all Americans, despite differences in tolerances for dairy products, preferences for other calcium-rich foods and susceptibilities to osteoporosis, as well as the lack of scientific evidence of benefit from dairy products for members of racial minorities. In this regard, federal nutrition policies do not yet address the needs of all Americans.
>
> (Bertron et al. 1999: 156)

In this case, recognition of significant population variation in lactase persistence and relatively high frequencies of impersistence among minorities in the U.S. warrants rethinking the explicit national policy of encouraging—indeed mandating—dairy product consumption by all. Those with different digestive biologies are being forced to conform to European-derived norms of dietary behavior, which the authors contend are associated with increased risk for various diseases such as osteoporosis and ovarian cancer, among others. In 2005 the PCRM filed a class-action lawsuit against grocery stores and dairies in the Maryland and Washington D.C. area, calling for milk carton labeling warning consumers that milk can cause serious digestive illness (see Figure 2.4). This charge echoed a 1979 lawsuit in which the Federal Trade Commission sued the California Milk Producers Advisory Board for its advertising campaign "Everybody needs milk." However, the judge ruled that there was insufficient evidence that milk was a significant threat to individuals with lactose intolerance, arguing further that:

> Milk is one of the most nutritious foods in the nation's diet, and from the standpoint of the population as a whole, or even significant population groups, is literally "essential, necessary and needed." The withdrawal of milk from any major population group would amount to a nutritional disaster.
>
> (Katz 1981: 267)

A response to the PCRM article appeared in the same journal three years later in a paper titled "Overcoming the Barrier of Lactose Intolerance to Reduce Health Disparities" authored by NDC researchers (Jarvis and Miller 2002). In this counterclaim, the authors argued that the relatively high rates of lactose maldigestion, and the concomitant low milk intake among minority populations, are significant contributors to their higher rates of chronic diseases such as osteoporosis, hypertension, stroke, and colon cancer. The basis of the claim is evidence suggesting that calcium and "other dairy-related nutrients" may reduce the risk of these diseases, although at present most studies presenting such results have relied on correlations and retrospective data, rather than demonstrating a direct causal link between milk consumption and lower risk

Figure 2.4 **PCRM's notice of their lawsuit against grocery stores promoting milk to minorities**
(Source: Physicians Committee for Responsible Medicine, used with permission).

of these diseases. Thus, to reduce disparities between the relative health advantages enjoyed by European-Americans, "Physicians can help reduce the disease burden and health care costs in minority populations by committing themselves to helping their clients overcome the barrier of lactose intolerance" (Jarvis and Miller 2002: 64). This is to be accomplished by providing such clients with "several simple strategies that allow those with low lactase activity to consume dairy products," as outlined previously.

In a separate publication the NDC argues that although "many minorities have low levels of lactase ... stereotyping all minorities as lactose intolerant is inappropriate" (National Dairy Council 2003b). Thus the claim of bias against minorities evident in recommendations to consume milk is turned on its head to suggest that individuals who fail to consume milk owing to fears of symptoms (often misplaced, the NDC researchers claim) from lactose maldigestion are at risk of major chronic diseases and well-known health deficits due to their dietary choices, and to accuse the authors from the PCRM of racial stereotyping. The NDC materials explicitly confirm that the Dietary Guidelines are for *all* Americans, that dairy foods are required to provide nutrients not found in other types of foods, and that minorities are especially at risk of calcium-deficiency diseases as a function of lower dairy consumption (National Dairy Council 2003b).

At the heart of this debate are two key issues. One is the nature and significance of biological diversity in lactase production in adulthood and its relationship to milk consumption. The second is the necessity of milk in the diet of contemporary U.S. citizens, a topic to be considered in more detail in Chapter 4. By and large, aside from the allegations of the anti-milk groups, the NDC and the USDA constitute the main voice in shaping the dominant rhetoric about biological diversity in lactase production. Their story is constructed around the inherent goodness of milk and the benefits of a biological make-up that allows for daily consumption of abundant quantities. Thus the older advertising slogan "Drink Milk for Fitness" had a perhaps unintentional link to evolutionary explanations, but of course it went further by making this a blanket statement which suggested that everyone's physical and perhaps Darwinian fitness might be enhanced by milk consumption.

This idea that people who drank milk were "better" than others dates back to the early twentieth century. As Samuel Crumbine and James Tobey wrote in their ode to the virtues of milk (1929: 77–78):

Throughout the course of history, milk has been hailed as the ideal food, an opinion which is well justified by our modern knowledge of the science of nutrition. The races which have always subsisted on liberal milk diets are the ones who have made history and who have contributed the most to the advancement of civilization. As was well said by Herbert Hoover in an address on the milk industry delivered before the World's Dairy Congress in 1923, "Upon this industry, more than any other of the food industries, depends not alone the problem of public health, but there depends upon it the very growth and virility of the white races."

This perspective was on full display just a few years later at the 1933 to 1934 World's Fair in Chicago. Sponsored by the USDA, a large installation in the 15,000-square-foot Dairy building emphasized the national as well as personal benefits of milk consumption with its slogan "Dairy Products Build Superior People." Among the exhibits was one extolling dairy's contributions to "Physical Perfection," with claims such as personal charm and beauty, a slim, lithe body, strong white teeth, beautiful hair and keenness of spirit for women, and coordination, rhythm, and strength for men (White 2009).

Not only was drinking milk and consuming other dairy products seen as superior, it was viewed as normal and normative. This perspective endured through the 1960s when studies began to demonstrate population variation in lactase persistence, with the modal global form being lactase impersistence. However, despite widespread recognition of this distribution, lactase impersistence remains implicitly pathological, as evidenced in widely used terminologies lactase *deficiency* and lactose *mal*digestion. Such medicalizing of non-Western, particularly African biologies appears in the colonial period, and remains evident in the example of lactase impersistence (cf. Comaroff and Comaroff 1992; Gould 1981). To be fair, lactose intolerance, which may result from lactase impersistence, is a cluster of uncomfortable physiological symptoms for which a person might seek medical help, but it is important to acknowledge that these only manifest in the context of milk consumption.

From the USDA's and NDC's perspective, avoidance of milk is not an acceptable strategy, regardless of one's lactase status. To consume milk in the United States is to be healthy; to avoid milk is to put oneself at risk of a variety of serious ailments. The NDC retains a vision of lactase impersistence as problematic, not so much because it may provoke gastrointestinal symptoms but rather because it may result in reduced milk consumption. The very term maldigestion suggests a malady, and the following passage describes the deviant nature of this condition: "Data from most studies suggest that individuals with primary lactose *deficiency* consume less milk than those who digest milk *normally*" (Jarvis and Miller 2002: 58, emphasis added). Thus, while they acknowledge underlying biological variations, it is of no practical significance; it should not be a barrier to consuming milk and enjoying the health benefits this confers on those who have a history of drinking milk and continue to do so regularly. One can and should "overcome" this biological deficit to achieve full participation in U.S.

culinary culture and its self-evident salutary consequences. A more recent article went so far as to claim that reports of high rates of lactose intolerance among African-Americans were largely a "myth" (Byers and Savaiano 2005), though it is curious why they targeted African-Americans and left out Asian-Americans, Native Americans, or Hispanics, populations with high rates of lactase impersistence and reported lactose intolerance.

Furthermore, because biological variation is discounted, the NDC suggests that it is negative cultural attitudes about milk that reduce its consumption by minority groups. Arthur Whaley noted this same trend in epidemiological studies: "Ethnic/racial groups are often seen as having misperceptions and unhealthy behaviors learned through cultural socialization that increase their risk for adverse health outcomes" (Whaley 2003: 738). Hence, individuals in such groups should be educated about the value of milk in their diets and enact behavioral changes to achieve a "better" diet; the source, value, or integrity of diverse "cultural attitudes" and culinary traditions is dismissed. Milk is deeply entrenched in U.S. culinary culture, national identity, and agricultural economy, and drinking milk is no less than full enculturation into U.S. life.

Yet, the increasing diversity of the U.S. public must be acknowledged and celebrated to some extent, especially if the goal is to sell more milk and reverse a downward trend in milk consumption. Thus efforts to embrace that diversity while simultaneously unifying it into a common milk-drinking experience have escalated.[9] It is impossible to avoid the images in popular culture from the wildly successful "got milk?" advertising campaign, which feature famous role models of various ethnic backgrounds sporting milk mustaches.[10] This practice is not lost on anti-milk groups, who complain that this is misleading owing to the relatively high frequency of lactase impersistence/lactose intolerance among groups of African, Asian, or Latin-American descent (and who have also subverted the "got milk?" slogan in various ways; see www.milksucks.com). By using minority role models, the dairy industry hopes to instill a positive association with milk and encourage milk consumption among minority children. Given that food preferences are established in childhood (Rozin 1983, 1990), the hope is that this strategy will pay off in the long term.

So how should food policies that take into account population variation in adult lactase production be constructed? Both the NDC and PCRM have valid claims—that to characterize minorities as lactase impersistent (or lactose intolerant) is to engage in racial stereotyping and that to mandate milk consumption for all U.S. citizens is discriminatory against those who are lactase impersistent (and especially those with lactose intolerance). To some extent the claims and counterclaims about the healthiness of milk are irrelevant to this discussion, although it is a travesty to attribute health deficits among minorities to their "failure" to drink milk. Milk is neither the elixir of life, whose consumption will surely prevent chronic disease, nor is biology destiny—drinking moderate amounts of milk is not likely to be seriously problematic for most people with lactase impersistence. Various lactose-reduced options exist for those who wish to consume dairy, but at the same time an increasingly diverse U.S. public is being led to believe that they must consume milk to be healthy, and traditional cuisines and alternate sources of milk's nutrients—especially calcium—are largely discredited.

In sum, dsespite widespread acknowledgment that a substantial minority of people in the United States—and the majority in the world—are lactase impersistent as adults, it appears that the strong cultural value placed on cows' milk and governmental support of the dairy industry inhibit policies that put the anthropological understanding of lactase persistence into practice. Thus while the latter emphasizes biological variation in milk digestive physiology and the unique historical processes that produced it, this perspective has been subsumed into subtly disguised normalizing discourses that downplay the significance of this diversity, and promote a modal biological response to milk that should facilitate its consumption by all U.S. citizens throughout life. The biology that facilitates ongoing milk consumption is likewise celebrated; other digestive physiologies are seen as abnormal or deficient in some way and need to be "overcome."

3
A BRIEF HISTORY OF MILK CONSUMPTION

Europe and the U.S.

The evolution of lactase persistence is closely tied to animal domestication and the use of domesticated mammals for milk. Dairying became an important part of subsistence practices only in the Old World, and as we saw in the previous chapter, lactase persistence is also unique to populations with ancestry there. Within the Old World, dairying was not ubiquitous, and lactase persistence evolved only among populations that made use of mammalian milk after weaning. Along with genetic adaptations to milk, the cultural valuation of dairy animals and dairy products became core components of dairying societies.

In this chapter I review the evidence for the origins of dairying, milk and other dairy product usage, and the emergence of dairying "cultures." The focus will be on Europe, where this history is best documented, and for the post-colonial period, the United States. As we will see, although European countries and countries populated by Europeans (the U.S., Canada, Australia, New Zealand) consume the most fresh milk in the world today, that was not typical of their history, since other forms of dairy products were more commonly eaten. I will trace the history of milk usage and ideas about the goodness of milk, particularly for children. The transition from dairying as a domestic enterprise to one characterized by industrial production and mass distribution is a defining moment in this history, as it set the stage for a new pattern of widespread consumption of fresh "sweet" milk.

Early Domestication of Mammals and Dairying

There are two major domesticated cattle species: European cattle (*Bos taurus*) and Indian cattle (*Bos indicus*). *Bos taurus* (Figure 1.4) were initially domesticated in the Fertile Crescent in the Middle East around 8,000 years ago, while the earliest evidence of *Bos indicus* (Figure 3.1) domestication is about 7,000 years ago in contemporary Pakistan. There may have been a third—and potentially earlier—domestication event for cattle in Africa. Well prior to domestication, European and Indian *Bos* lineages had already separated (Troy et al. 2001). Based on genetic evidence, European cattle appear to have derived from Near

Figure 3.1 **Bos indicus** (Source: Scott Bauer, United States Department of Agriculture).

Eastern progenitors (wild oxen, or auruchs) sometime between 37,600 and 10,100 years ago (Troy et al. 2001).

Current evidence suggests that milk usage was part and parcel of the origins of the larger agricultural complex—the domestication and cultivation of plants and animals—in the near East and Europe. Based on the presence of milk residues in ceramic vessels, it appears that milk utilization in Europe originated in the Balkans of Central Europe around 7,900 to 7,500 years ago (Craig et al. 2005). It subsequently spread North and West, and was evident in Britain by 6,100 years ago (Copley et al. 2003). This range of dates coincides with the emergence of both cereal agriculture and animal domestication in Europe, suggesting that dairying was a part of the original suite of subsistence practices that spread through Europe during the Neolithic era. Some researchers have suggested that dairying may have been a key reason for the domestication of cattle, sheep, and goats in the Near East and its subsequent extension into the Mediterranean region (Vigne and Helmer 2007). While these animals have many uses for early agriculturalists including meat, skins, wool, or traction, it now appears that milk production was among the very first uses.

In western Turkey, between 9,000 and 7,000 years ago cattle were more common than goats and sheep at sites where pottery shards also indicate milk processing (Evershed et al. 2008). Subsequently, cattle become the predominant species in Central and Northern Europe and dairying became an entrenched component of European agricultural economies. In the Mediterranean, sheep and goats were more common, but also appear to have been used primarily for products other than milk. In part this is due to ecological differences, as cows thrive in cooler, wet grasslands, while goats and sheep tolerate dry climates and the rugged terrain of the Mediterranean. In this region milk was processed into various forms of yogurt, cheese and other milk-derived products rather than being consumed fresh. The agricultural differences between Southern Europe, the Mediterranean, and Central and Northern Europe, as well as historic reports and the lower frequency of the allele for lactase persistence, point to this. For instance, the Romans used goats' and sheeps' milk for the production of cheese, while cattle were used primarily as a draught animal. In contrast the inhabitants of Central and Northern Europe practiced cattle dairying, and it has been suggested that they drank milk—or some fluid form of dairy products—in significant amounts. As Strabo, the Greek geographer and historian, described the "men of Britain" in his famous *Geography*, "Their manners are in part like those of the Celti, though in part more simple and barbarous; insomuch that some of them, though possessing plenty of milk, have not skill enough to make cheese, and are totally unacquainted with horticulture and other matters of husbandry" (quoted in Lee 1900).

The co-evolution of European lactase persistence with dairying may have occurred in the following scenario. Between 8,000 and 6,500 years ago, during the early Neolithic cultures of Eastern-Central Europe where cattle dairying had become an increasingly important part of subsistence, there was positive selection on the individuals who had the mutation for lactase persistence. Analysis of ancient DNA extracted from bones from five archaeological sites from Northern Europe that date up to 8,000 years ago showed that the common European allele for lactase persistence was not present prior to that time (Burger et al. 2007). After the emergence of what is known as the LBK culture (linear-bandkeramik or linear band pottery culture, identified by a distinctive banded decorative element used on pottery; LBK groups are generally considered to be responsible for major agricultural innovations in Central Europe), lactase persistence frequencies rose rapidly in a co-evolutionary process. The allele for lactase persistence spread through the demographic expansion that ultimately resulted in the establishment of highly developed cattle-based dairying economies during the Middle Neolithic of Central Europe around 6,500 years ago (Beja-Pereira et al. 2003). Beja-Pereira and colleagues (2003) also found a strong geographic overlap between regions of cattle use, high frequencies of lactase persistence, and variability in several milk protein genes among local cattle populations indicating selection for specific milk protein profiles.

Pre-industrial Milk Production and Consumption

The extent to which fresh milk was consumed in Europe prior to the 19th century is not known with any certainty. Most reports suggest that it was rare; fermented milk products were more common in the diet, and milk was prized in large part as a source of fat rather than as a fluid. Various forms of "soured" milk such as buttermilk (the liquid remaining after the cream is taken off and made into butter), yogurt, and various forms of cheese were most commonly consumed. "Fresh" milk would have been difficult to maintain for any length of time, especially during the summer months when it was warm and cows were "in milk." Thus the only context in which it would have been consumed fresh was shortly after milking. Milk and whey were commonly used in grain-based porridges or soups in Northern Europe (Mendelson 2008). From the perspective of lactase persistence, fresh milk, whey, unfermented buttermilk, or fresh soft cheeses made from these products would have been the only major sources of exposure to lactose, but given the ubiquity of this novel trait in Northern European populations, these must have been important dietary components.

It does appear that mammalian milk was used as a weaning food during the Roman period, and ongoing exposure to milk in the post-nursing period may have been critical to the spread of lactase persistence. Using the presence of nitrogen and carbon isotopes in a skeletal sample from Egypt (ca. 250 AD), researchers have been able to infer that starting at around six months of age, infants were supplemented with animal food products, most likely the milk of goats or cows (Dupras et al. 2001). These data are consistent with child-feeding recommendations in Roman medical texts by Galen and Soranus, which advise supplementation with goats' milk and honey (which unfortunately increases the risk of anemia [goat milk is low in iron] and botulism [from *Clostridium* bacteria in honey]). Interestingly, in the Talmudic literature there are also recommendations for infants and children to consume honey, but nuts were identified as the quintessential children's food (Weingarten 2005).

There are a number of other sources of insight into child diets in European history. Depictions of children with food in art or literature are one. In paintings from the Renaissance period and later, children are often painted with fruits such as apples or grapes. These are most likely not to be read literally as important elements of the diet so much as they are symbolic of fertility, or in the case of the apple, temptation, or the source of knowledge or virtue (Riley 2004). Infants are often seen nursing but, importantly, when children are painted in association with foods, a glass of milk is not among them. In family portraits around well-laden tables of food, wine is featured as the only drink. Medical treatises from the Middle Ages sometimes recommended diluted wine for infants and young children, along with water and honey (Adamson 2004). In the Netherlands, fresh milk was not considered a healthy food, resulting in gastrointestinal or circulatory troubles and headaches. It was also considered bad for the teeth (van Winter 1992).

Other data sources are dietary budgets for households, classes of workers, orphanages, or poorhouses, whose rations were prescribed by municipalities or employers. In the mid-19th century, records from English poorhouses indicate that children aged between 2 and 9 years were given bread, porridge (made with water or milk), rice pudding, pea soup, meat, potatoes, and cheese. Sometimes they were given beer or tea (Alcock 2004). There was wide variation in the rations given to children; in some institutions children were provided with up to a pint of fresh milk or buttermilk per day, less for older children. Thus there was some recognition that milk might be especially beneficial as a food for growing children—especially young children.

Overall, it is safe to say that the diet of most Europeans during the Middle Ages was monotonous, with bread (or porridge in the British Isles) being the staple food, with small amounts of meat, lard, butter or cheese as a means to "make the bread go down" (Bernard 1975: 35). Food shortages and more severe famines due to crop failures and wars were also fairly common. After the Black Plague, which terrorized much of Europe during the 14th century and caused the death of about one-third of the population, there was a period of relative prosperity. This correlated with improvements in the quality of the diet, with increases in the consumption of animal products, including dairy (Teuteberg 1991). However, as populations rebounded and urban areas expanded, the diet again reverted to one based heavily on carbohydrates. Around 1800, bread was the core dietary staple, supplemented with modest amounts of meat (beef, mutton), cheese, and beer. In Ireland, butter and cheese had been staple parts of the diet, with grains grown more for beer production than for bread, but, with increased emphasis on grain production for export and potato cultivation in the 18th century, dairy product consumption declined, and by 1800 rural peasants were unlikely to own any cows (Cullen 1992). In Belgium in the early 19th century, dairy products in some form or another contributed around 300 kcal to the diet, representing about 15% of total calories (Bekaert 1991).

Prices of agricultural commodities and wages from 1500 to 1800 also give us some sense of the role of milk in the diets of pre-modern Europeans. During this time prices rose and wages declined, and it has been well documented in England in particular that households adjusted their budgets in response, with declining consumption of meat and especially dairy products. Furthermore, with the Enclosure Acts of the 17th and 18th centuries, which restricted rural households' access to common grazing areas, fewer households owned cattle and had access to milk (Shammas 1990). Thus the price of dairy and other animal products rose, while that of other commodities including those pro-duced in European colonies was declining. The consumption of both milk and cheese fell, while there was an increase in butter usage, probably as workers switched to bread from porridge, taking it as a meal with tea or coffee.

As urban centers grew in Europe, large populations lost direct access to milk, instead relying on purveyors who brought milk or other dairy products in from rural areas. Needless to say, fresh milk would likely have begun to ferment during its journey to and distribution within cities. Urbanites throughout Europe, especially the poor, likely had virtually no access to fresh milk (den Hartog 2001). In Britain in the late 18th century the rural poor consumed an average of about 2.5 ounces of milk per day; 40 or 50 years later, poor urban workers consumed 1.9 ounces (Clark et al. 1995), although we do not know whether this was fresh milk or some form of fermented milk such as buttermilk. In comparison, in late 18th- and early 19th-century Vienna, a relatively wealthy city, annual per capita intake of milk was around 40 liters (~10 gallons; this works out to about 3.5 ounces [< ½ cup] of milk per day) (Sandgruber 1992). For the urban working class in Vienna, however, milk was not part of the diet. Cheese was the only dairy product consumed with any regularity.

Dairy in Switzerland

Since milk is often thought of as an especially central part of the diet in Switzerland (due in no small part to it being home to Nestlé, a major global manufacturer and supplier of milk products), it is worth taking a closer look at the role of milk and dairy in the diet there. Due to the mountainous terrain of the Alps, which makes agriculture difficult, and the existence of alpine pastures, animal husbandry and dairy farming in particular was the predominant mode of subsistence. Barbara Orland's historical analysis of dairy farming in Switzerland (Orland 2004) provides insight not only into the complexities of milk production there, but also into the emergence of Switzerland as a major producer of manufactured dairy products such as cheese.

In the upland pastures, the lone mountain hut, made of wood or stones, housed alpine herdsmen or dairymaids who occupied it for several months of the year (see Figure 3.2). Milk was processed there into butter and cheese, the most important commodities made from milk, and there was a widespread belief that the best milk came from these alpine areas, in part because the fat content of the milk was much higher and hence the yield of butter and cheese was greater. The remaining skimmed milk could be combined with whey and converted into a low-fat cheese, although this had a much shorter shelf life than harder high-fat cheeses. While there were local trade networks for raw milk and processed dairy products, there was no commercial cheese industry until the 19th century. With efforts to increase "efficiency" of milk production and transport, dairying activities became more focused in the lowlands, with cows being fed locally grown fodder instead of grazing in upland meadows, and by the mid-19th century the alpine system of dairy farming was disappearing.

Figure 3.2 **The Swiss alpine dairy system, as illustrated in a print from the late 19th century** (Source: Print from ca. 1890).

Ironically, however, the image of the alpine herder was transformed into a nationalist symbol. The Swiss Alpine Farming Association, founded in 1863, praised activity in the pure alpine air and a diet of cheese, butter, milk, and meat as a source of health, physical strength, and the spirit of freedom of the Swiss people. The shepherd embodied these Swiss virtues and a variety of milk symbols became important nationalist icons. Although milk had long been produced in Switzerland, its primary products were butter and cheese, with the latter predominating with restrictions on the butter trade. However, as with other sites in Europe and the United States, it was not until the late 19th century that fresh milk was sold for the purpose of human consumption. It is also worth noting that while Switzerland has become known for its delicious chocolates, "milk chocolate" was a byproduct of innovations in milk-processing technology (condensed and powdered milk). With public health rhetoric about the value of milk to children's diets increasing (see the following section), the addition of milk to chocolate made it a suitable—even advisable—product for children's diets (Chiapparino 1995). The fact that Switzerland produced an abundance of milk also made it possible for them to expand their chocolate industries, which were established in sites of traditional cheese production, where milk-processing equipment was readily available.

We may reasonably conclude that fresh milk consumption was neither the norm nor normative up until the mid- to late 19th century in Europe (Fenton 1992; van Winter 1992: 999, see other contributions in Lysaght 1992). The lack of cold storage, especially during the summer months of high milk production, prevented widespread distribution beyond the household, or consumption long after the actual time of milking. More frequently, milk was processed into cheese and butter, which could be stored and transported without becoming contaminated. Milk was thus more often consumed as a food than as a drink, and was especially valued as a source of butterfat. The liquid portion—whey and buttermilk—was used as the base for porridges or turned into softer cheeses. These products would have been more readily available to rural consumers, but as land was turned over to grain production, rural sources also became scarce (Den Hartog 1992). Beer and wine were more common drinks, with tea and coffee (drunk with some milk and also sugar) added during the colonial period. That said, we really have no clear documentation of the extent of fresh milk consumption prior to the late 19th century. Undoubtedly it was consumed, but how much, and who consumed it is not known with any precision. Based on poorhouse records it seems that milk was preferentially given to young children, but how extensive this pattern was, and whether children were viewed as important consumers of fresh milk remains unclear.

In North America, there was little animal domestication and no dairying prior to the 16th and 17th centuries when Europeans began to establish colonies there. Northern Europeans were the earliest settlers, and they brought their domesticates with them and established the agricultural and culinary systems that were familiar to them from home (Crosby 1986). These included dairying, especially cheese and butter making. Recipes from the early period include those for making different types of hard and soft cheeses, milk-based puddings, clotted cream, cheesecakes, and various kinds of pancakes (Eden 2006). In all likelihood the colonists had access to more cows and dairy products than did their European counterparts, as there were fewer constraints on land and thus easy availability of grazing areas for cows (Shammas 1990). Commonly consumed beverages included coffee with cream or milk (and if available, butter and a whole egg!), hot chocolate (also with milk), tea, or cider (fermented to varying degrees). Thus fresh milk was used, but in association with other ingredients in hot beverages, while buttermilk would have been the main way in which a fluid form of milk was consumed on its own. Cheese making remained the main way in which milk was processed for consumption. It was an important part of a household's domestic economy, and overseen primarily by women.

Urbanization and the Rise of Fresh Milk Consumption

Although contemporary images of milk production tend to highlight the bucolic conditions of cows grazing in verdant meadows, the trend toward fresh milk consumption began amidst the grit and grime of 19th-century cities in Northern Europe and North America. To be sure, there is some evidence that fresh milk consumption had been rising prior to this time, including among rural populations, but the transition was marked. For example, in England, poor workers in 1863 consumed 3.7 ounces of milk per day (only ~ ½ cup), which was almost double that of a mere 25 years earlier (Clark et al. 1995). In Vienna, average consumption almost quadrupled from 1850 to 1910 (Sandgruber 1992: 986).

The question is thus: Why did urban populations in particular begin to consume larger quantities of fresh milk, a form of dairy consumption that had previously been rare? Second: How could milk be made available to urbanites when it is highly perishable and vulnerable to spoilage by bacteria, given that this was likely one of the major problems associated with fresh milk consumption prior to this transition?

Anne Mendelson (2008) describes the "perfect storm" that produced the explosion of fresh milk drinking among urbanites. As rural migrants moved into cities in search of employment due to rural land reforms and opportunities for urban factory jobs, they had to rely on retail suppliers for all their food. These suppliers in turn required sources for large volumes of food commodities that could be easily brought to the urban market. Proximity was especially important for fresh foods such as milk, meat, or vegetables, as prolonged time in transport would lead to spoilage. In addition, consumption of novel kinds of beverages such as coffee or tea that were produced in the colonies skyrocketed across socio-economic groups including the working class (who used the caffeine and sugar to maintain their long working hours: Mintz 1985), and these were drunk with milk. Furthermore, without the domestic technologies of rural households, knowledge, or time to process milk into traditional forms, urbanites became the ideal market for fresh milk.

Another change was also in play. There was growing scientific recognition of the process of "putrefaction," which in the case of milk led to the familiar tangy, acidic taste of fermented buttermilk or the sourness of yeast-fermented bread. While the mechanisms underlying this process were not understood at the time, it was proclaimed undesirable as a cause of disease or digestive problems. This meshed seamlessly with Louis Pasteur's experiments with wine fermentation ("spoilage") followed by Robert Koch's germ theory of disease in the later 19th century. Thus authorities concerned with public health proclaimed that milk was best consumed fresh, and bread, which had long been leavened through prolonged yeast fermentation, was to be chemically (and quickly!) leavened with sodium bicarbonate.

Not only did these changes stimulate growth in the demand for and supply of fresh milk, but working-class women were entering the industrial workforce to help support their families. Their long factory shifts precluded nursing their infants, and a substitute for breast milk was to be found in fluid cows' milk. Urban middle-class women had become isolated from rural community networks that supported breastfeeding, and cows' milk became the replacement for wet nurses. This downward trend in nursing helped create the market for commercially produced cows' milk and underpinned campaigns that highlighted milk as the ideal food/drink for children (DuPuis 2002). However, before it could be firmly established in the public's mind that children—from infants to adolescents—needed to drink cows' milk, profound problems of milk safety had to be solved.

In response to the demand for milk by urban residents, peri-urban dairies sprang up. Cows were densely packed into stables, which were often attached to breweries and distilleries, and sustained on the fermented grain mash byproduct of beer and liquor production, so-called "swill." Not surprisingly, cows housed under such conditions were more often than not diseased, and their milk rife with pathogenic bacteria (see Figure 3.3). Several species of bacteria thrive in milk including *E. coli*, *Salmonella*, *Mycobacterium* (cause of tuberculosis), *Brucella* (cause of brucellosis or undulant fever), and *Lysteria*, among others,

Figure 3.3 **A ghoulish figure selling swill milk to a mother and her sickly child in New York City** (Source: Harpers Weekly (1878)).

which were significant causes of illness in urban milk consumers. Milk from the 19th-century peri-urban dairies has been referred to as "the white poison" (Atkins 1992). Infants fed such milk died at an astonishingly high rate. Infant mortality rates in northeastern U.S. cities were around 50%, and swill milk consumption, in the context of poor sanitation, crowding, and poverty, contributed to these high rates. Physicians recognized that cows' milk-fed infants had higher mortality than breastfed or wet-nurse-fed infants, but had no means of understanding the cause of this difference.

Not surprisingly there were calls for reform in milk production. One early proponent was Robert Hartley, an evangelical social reformist who targeted his energies at alcohol consumption and the swill milk system upon which New York City residents depended. Hartley's argument for changes to this system relied heavily on assumptions that milk drinking—and by that he meant cows' milk rather than breast milk—was a natural state of affairs for children in particular. In his treatise *An Historical, Scientific, and Practical Essay on Milk as an Article of Human Sustenance* published in 1842, Hartley stated that milk is an "essential part of human sustenance" and "it is the best and most palatable aliment for the young; it is suited to nearly every variety of temperament and is adapted to the nourishment of the body in every age and condition" (Hartley 1977[1842]: 75). Hartley's voice was joined by that of John Mullaly, who in 1858 described milk as a "principle article of food of all children, and when it is impure it is not reasonable to suppose that they can be healthy" (Mullaly 1853: 23; see also Figure 3.4).

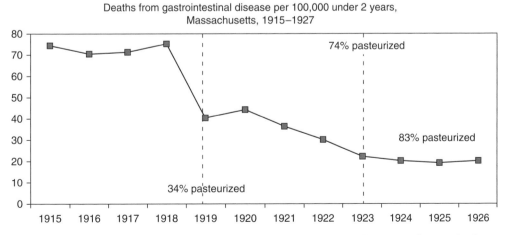

Figure 3.4 **Correlation of trends in child deaths from diarrheal disease and spread of pasteurization in Massachusetts** (Source: Drawn from data in Crumbine and Tobey (1929)).

While both Mullaly and Hartley brought the contamination of the milk supply and its deadly effects on children to public attention, both worked from the assumption that children should be drinking milk, and that cows' milk was "naturally" a food needed by them; that is, milk was seen as inherently good, pure, and natural. Milk's white color, maternal associations, and Old Testament odes to a land that "floweth with milk and honey" were deployed in support of this view.

It remains unclear where this assumption came from. It may have long been latent, part of an intuitive recognition that since infants begin life consuming milk, milk provides the resources to support growth and that these could be easily extended to post-weaning-age children. Furthermore, since infants needed milk every day so should older children consume it that frequently. Whatever its origin, the dialogue about children and milk was framed not in terms of whether children (that is, post-weaning-age children) needed milk, but rather in terms of the importance of milk hygiene. This is not surprising given that the mid- to late 19th century was a period of sanitary reform, and reform of the milk production system was part and parcel of this larger trend in public health. However, it may also be that contaminated milk became a convenient way in which to frame entrenched problems of urban poverty and disease (a situation not unlike that described in the previous chapter wherein the National Dairy Council attributed higher rates of chronic diseases among minorities to their lack of milk consumption). More tractable than those larger social problems, it provided a convenient means by which a solution could be envisioned.

Hartley also side-stepped the question of whether infants would be better off with breast milk or cows' milk by focusing on the harm done by feeding them the *impure* cows milk produced by the swill-fed cows of peri-urban dairies:

> As might be expected, the cattle, under this most unnatural management, become diseased, and the lactescent secretions not only partake of the same nature, but also are impure, unhealthy, and innutritious. Yet this milk is the chief aliment of children in all places where the population is condensed in great numbers; it is the nourishment chosen and relied upon to develop the physical powers and impart vigor to the constitution during the most feeble and critical period of human life, when the best possible nourishment is especially necessary, in order to counteract the injurious effects of the infected air and deficient exercise, which are often inseparable from the conditions of city life.

(Hartley 1977[1842]: 109)

Further,

> They force the milch cows with swill so that they literally become drunkards, and send the milk to rear our children, as you do in New York,—thus sowing the seed of disease in the cradle, and poisoning the fountain of life at its source.

(Hartley 1977[1842]: 239)

Regulating the Milk Supply

To remedy this problem, Hartley and Mullaly recommended extensive inspection of the retail milk supply and the closure of swill-milk peri-urban dairies, and encouraged countryside dairies to produce for the urban market. This required improved methods of transportation to ensure a fresh and safe supply of milk, and expansion of the rail system in the 1850s enabled this process. However, even though time to market was shortened by the rail system, milk is a highly perishable product, and rail transport in no way guaranteed uncontaminated milk. The bacterial sources of that contamination were revealed with the work of Robert Koch, founder of the germ theory of disease (he discovered the bacterial cause of tuberculosis in 1882), and Louis Pasteur, from whom the term pasteurization derives (in 1862 he demonstrated that heating liquids such as milk destroyed the microbes that induced fermentation, or spoilage).

There are different types of pasteurization, but in the late 19th and early 20th century, the practice involved heating milk to 145 degrees F for 30 minutes (Mendelson 2008).[1] This is well below the boiling point, because at higher temperatures milk curdles (the casein proteins aggregate, causing a separation of solids and liquid). This was sufficient to kill off most microbes in milk. However, even pasteurized milk requires cold storage to prevent microbial expansion, and it was the combination of pasteurization and refrigerated rail cars, first patented in 1867 and in wider usage in the late 1870s, that ultimately made mass production and distribution of fresh milk possible. Indeed refrigeration made possible the valorization of "fresh" food, especially among a growing urban populace long (both in distance and time) displaced from traditional agricultural loci of production (Freidberg 2009). Some scholars have suggested that it was really refrigeration that contributed most to improvements in 19th- and 20th-century diets, particularly because it allowed more convenient access to animal products such as milk (Craig et al. 2004). Craig and colleagues suggest that refrigeration alone accounted for a 1.7% increase in annual dairy intake and thereby contributed to the increase in height seen in the early 20th century (see Chapter 4 for more discussion of the milk-height relationship).

Despite the known benefits of pasteurization—especially when cows were crowded together in unhygienic conditions—there was resistance to mandating this technology. In part this stemmed from urban physicians who supported a milk certification program, wherein a medical board would "certify" a dairy based on its sanitation and inspection of its cows and milking practices (DuPuis 2002). Thus milk would be sold in its raw state but would be certified for safety. Certification was a labor-intensive process, especially given the increased number of dairies in the late 19th century, and the cost of testing cows for pathogens would have fallen to dairy farmers. Certification supporters were concerned about pasteurization's negative impact on the "vital" components of milk, and also worried about pasteurization as a means by which "dirty" dairy farms would persist, as any contaminants would be destroyed in the pasteurization process. In the end, pasteurization of the milk supply became the path of least resistance, but it was not mandated by any municipality until 1908, when Chicago became the first city to require it (Chicago also had the dubious distinction of allowing swill dairies until 1892, long after other major cities had banned them). The first state-level mandate did not occur until 1947 in Michigan. State laws rather than federal laws still regulate the dairy industry, although in practice the 1924 U.S. Public Health Service's *Standard Milk Ordinance*, which outlined standards for milk processing (including pasteurization), packaging, and selling, were widely adopted. Any interstate commerce in dairy products is subject to these federal requirements.

Pasteurization was just one of several sanitation initiatives in the works in the late 19th and early 20th centuries. The recognition that microbes were in the water and food supply and the cause of many diseases led to a variety of regulatory efforts to assure a safer environment, particularly in cities. As a result it is difficult to ascertain precisely what kind of effect milk pasteurization had on child health and mortality. As Figure 3.4 demonstrates for Massachusetts, the spread of pasteurization was correlated with a decline in child mortality from gastrointestinal diseases, which are most commonly attributable to contaminated food or water. Importantly, however, long before pasteurization was adopted as the means of getting fresh milk to market, other industrial forms of milk preservation had been invented. Gail Borden, for example, developed a vacuum condenser mechanism for producing condensed milk and preserving it in metal tins, and his success was owed to the widespread use of his product by Union soldiers during the Civil War. To make condensed milk, the water from milk is extracted through a process of vacuum evaporation (which lowers the boiling point to room temperature) and sugar is added to form a thick, viscous paste. It was various forms of tinned milk (evaporated milk, which unlike condensed milk has no sugar added, was developed in the 1880s) that became the commercial products advertised and sold as safe infant foods (see Figure 3.5).

Figure 3.5 **Angelic babies promoting Borden's condensed milk** (Source: Borden Company *E pluribus unum: the story of an eagle* (1904). Courtesy of Smithsonian Institution Libraries, Washington, DC. Used with permission).

Because pasteurization and canning allow for longer storage of milk, the volume of milk production could increase, but dairies would also need facilities to accommodate this larger supply and its storage. The net effect was the merging of dairies into larger corporate units, with a centralized processing facility for pasteurization and market distribution. Processing thus became separated from production, with processors buying milk from multiple dairies. In response, there were incentives for milk producers to increase the scale of production to remain competitive pricewise. Milk, like other products such as bread, became an industrial, mass-produced, distributed, and marketed product. This occurred more or less simultaneously in the United States and Northern European countries, and for the first time in their history, access to fresh fluid milk became ubiquitous.

Marketing of Milk: Normative Discourse about Milk and Children

As the volume of milk production increased, distribution networks developed, and methods of preservation were more widely adopted, growing the market for milk became a goal of the emerging U.S. dairy industry. Commercial, non-governmental, and governmental agencies became involved in the promotion of milk, and worked together in various ways to establish and then solidify milk's essential role in the American diet. The growing size and economic strength of the dairy industry, now producing a glut of milk, led to the formation of the National Dairy Council. Established in Chicago in 1915, its goals were to support research on the healthfulness of dairy products, and educational campaigns and other kinds of promotions of milk and dairy products. The NDC established a close alliance with the United States Department of Agriculture (USDA), which President Lincoln had established in 1862. As the USDA goals were to both foster and promote production and consumption of U.S. agricultural commodities and the formulation of government dietary recommendations, NDC and government interests were aligned, and as we saw in Chapter 2, this close relationship has endured into the 21st century. Between public health promoters of milk in the 19th century, technologies to transport and distribute relatively safe fresh milk, and government support of the dairy industry, the NDC was poised to gain increasing authority over child health, particularly when it came to promoting an ideal diet for children. Child health came to be equated with milk drinking, and through NDC's efforts—combined with government programs, education campaigns, and clinical recommendations—it became increasing difficult to imagine how children of all ages could be healthy without milk; that is, cows' milk.

Given that fresh milk still had a high likelihood of contamination well into the 20th century, it was initially difficult to promote milk as an especially salubrious beverage, especially for children. Indeed the earliest advertisements were for tinned rather than fresh milk, and they highlighted the safety of their products. Most significantly, however, advertisements were to feature children consuming milk, rather than the pastoral images of milk production that had predominated before the Civil War (DuPuis 2002). It is not terribly surprising that milk promotions should target children. After all, the impetus for a growing dairy industry had come from the need to provide milk for urban infants. The earliest advertisements highlight babies—healthy, plump babies thriving on tinned milk of various kinds (see Figures 3.5 and 3.6). Babies in these ads are described as "healthy," "strong," "contented." After the turn of the century and throughout the 20th century, advertisements tended to feature toddlers and school-age children, and the milk they drink is no longer readily identifiable as from a tin, but rather fresh milk in a glass or bottle (Figure 3.7).

Milk's construction as an essential food for children was bolstered by discoveries in nutrition, which was a very young science in the early 20th century. With the germ theory of disease still new and seemingly of enormous explanatory value, there was initially resistance to claims that food components were responsible for syndromes such as pellagra (now known to be a niacin deficiency) or beri-beri (a thiamine deficiency). In 1912 Casimir Funk coined the term vitamin, meaning a substance that was "vital" to life. Shortly thereafter, Elmer McCollum discovered Vitamin A and B (now recognized as a complex of separate water-soluble vitamins) and named Vitamin C as the substance preventing scurvy. McCollum later named Vitamin D in 1922, after discovering it to be the key vitamin that prevented rickets. With regard to milk,

Figure 3.6 **An early 20th-century advertisement for Highland evaporated cream for infants** (Source: *Ladies Home Journal* (November 1902)).

McCollum and others were able to demonstrate that milk was rich in a variety of vitamins, and that these made rats and mice supplemented with milk grow at faster rates and to larger sizes than those that were not supplemented (see McCollum 1957).

As Harvey Levenstein noted in his book *Revolution at the Table*, "vitamins and minerals were an advertiser's dream" (Levenstein 1988: 152). They allowed marketers to be specific about the qualities of their products and to back up their claims by reference to scientific discoveries. Initially Vitamin A was thought to be milk's unique growth-promoting nutrient, although as the twentieth century wore on, calcium came to predominate as the most important of milk nutrients that might be otherwise limited in the American diet.

Figure 3.7 **Fresh cows' milk as a food for all children, and a normative aspect of children's diets** (Source: From the back cover of a Capital Dairy Co. cookbook, *More Milk in the Menu* (1928). Photo courtesy of Special Collections (Sacramento Room) of the Sacramento Public Library).

Early advertisements reference milk as a source of "mineral phosphates," then protein, riboflavin, and Vitamin A; later calcium, phosphorous and Vitamin D. Milk contains relatively little Vitamin D, but with the discovery that it prevented rickets (a condition in which the weight-bearing bones bend due to a lack of calcium matrix) by enabling calcium uptake, milk started to be fortified with Vitamin D in the 1930s. Advertisements highlight milk's calcium starting in the 1930s (see Figure 3.8), but these messages took off with more vigor in the 1960s. Calcium remains the most trumpeted nutrient, symbolizing the inherent goodness and superiority of milk as a food. More will be said about calcium in Chapter 4.

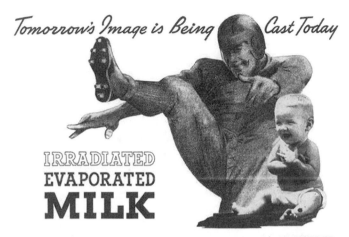

Tomorrow's Image is Being Cast Today

IRRADIATED
EVAPORATED
MILK

TEAMWORK (Calcium – Phosphorus – Vitamin D) IS NEEDED
TO ASSURE PERFECT BONE STRUCTURE!

Figure 3.8 **An advertisement from 1938 highlighting milk's calcium, phosphorous, and Vitamin D** (Source: *Journal of the American Dietetic Association* (November 1938)).

As knowledge about nutrition and the effects of nutrients on human biology grew in the 1910s and 1920s, there were parallel public health campaigns focused on child growth. Milk advertisements fit right in with these. Among early 20th-century advertisements were statements that milk "builds up the body—brain—bones—and muscles—and promotes healthy growth of the entire system" (Figure 3.9). These ads, combined with scientific findings of a positive effect of milk on growth, played well with a middle-class public that had become more child-centered, concerned with the special aspects of child health, and preoccupied with the physical growth and feeding of children (Levenstein 1988).

School health programs took up the charge to encourage milk consumption among schoolchildren. In various localities special school milk programs had been established in the 1920s. Students generally paid for these supplementary meals, but they would be subsidized by local charitable organizations such as the Red Cross or Anti-Tuberculosis Association for needy children (Crumbine and Tobey 1929). In a Newton MA school district, children who were 10% or more underweight were required to report for mid-morning milk. Moreover, children were often served milk in the classroom "while the teacher tells them about this wholesome product ... [and] ... to cultivate a taste for milk as the one best food for growth" (Crumbine and Tobey 1929: 155).

Schools became—and remain—a common venue for the promotion of milk, which occurred in school feeding programs and the curriculum. A pamphlet prepared by the Child Health Organization for the Bureau of Education reported on efforts to promote milk in primary schools and offered curricular

Figure 3.9 **Milk promotion emphasizing milk's value for child growth** (Source: Ladies Home Journal (June 1904)).

recommendations. Entitled "Milk and our School Children" and published in 1922, it provided evidence that half of a large sample of American schoolchildren were not drinking milk daily (Reaney 1922). Over one-third of children drank coffee, and almost 20% drank tea. The author concluded that "American parents and children must be taught the nutritional value of milk" (Reaney 1922: 4) "because milk is the best and most important food in the diet of the school child. No other food can take its place. It contains the elements necessary for the growth of the different structures of the body and provides heat and muscular activity [i.e. fuel, or calories]" (Reaney 1922: 8). To this end a variety of educational activities were proposed in which milk featured in every subject: "milk problems crept skillfully and beguilingly into every crack and cranny of the regular week" (Reaney 1922: 10). Accompanying these lessons, students

Figure 3.10 **U.S. Bureau of Education recommendations for schoolchildren** (Source: Reaney (1922)).

charted their daily milk consumption and their weight and height were also tracked.

While growth monitoring is no longer a central part of school nutrition programs, these lessons are remarkably similar to those currently proffered in nutrition education materials made available to schoolteachers from the National Dairy Council at the website www.nutritionexplorations.com. The milk education campaign in schools (Figure 3.10) directly corresponded to the NDC's recommendations for healthy children at the time, which were as follows. Drink four glasses of milk each day; eat some vegetables besides potato every day, eat fruit every day, drink at least four glasses of water every day, play part of every day out of doors, sleep many hours with the windows open, brush teeth every day, take a bath more often than once a week (Crumbine and Tobey 1929). While milk is recommended, these guidelines go well beyond this, indicating the authority that the NDC had over child health.

School milk programs were formalized as part of broader school feeding initiatives in 1946, with the National School Lunch Act:

It is hereby declared to be the policy of Congress, as a measure of national security, to safeguard the health and well-being of the Nation's children and to encourage the domestic consumption of nutritious agricultural commodities and other food, by assisting the States, through grants-in aid and other means, in providing an adequate supply of food and other facilities for the establishment, maintenance, operation and expansion of nonprofit school lunch programs.

(USDA Food and Nutrition Service 2009)

Fluid milk must be offered as part of school lunch programs in order to receive federal reimbursement, and years after the School Lunch Program was authorized, private institutions devoted to the care and training of children were also made eligible for milk reimbursements. The normative associations between milk and children of all ages became formalized in such government programs, and the cow—as the source of this essential food—was portrayed as the "foster mother of mankind" (White 2009).

Not Drinking Their Milk: Declining Milk Intake in the U.S.

Despite a plethora of support for milk in the U.S., consumption has been steadily declining since it peaked during World War II. As Figure 3.11 shows, per capita fluid milk intake in the U.S. averaged around 35 gallons per year (about 1.5 cups or 0.35 liters per day) in the 1920s and 1930s and reached its maximum at around 45 gallons per year during late World War II (almost one pint or 0.46 liters per day). The average American's milk intake today stands at 25 gallons per year, which works out to just over one cup per day (0.26 liters)

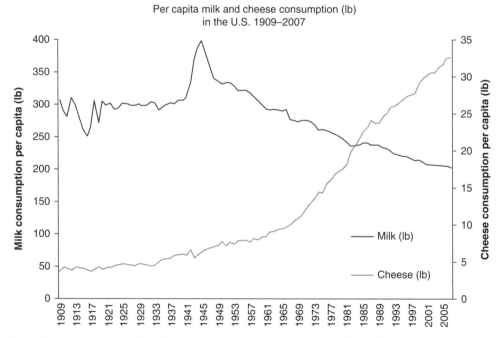

Figure 3.11 **Trends in fluid milk and cheese consumption in the United States** (Source: Drawn from data from United States Department of Agriculture (2008)).

(United States Department of Agriculture 2008). It is most likely that the dramatic upsurge during World War II reflects the rationing of many desirable foodstuffs such as meat, sugar, butter, and in some cases, cheese. Fresh milk was not rationed, and the production of milk increased to meet this demand.

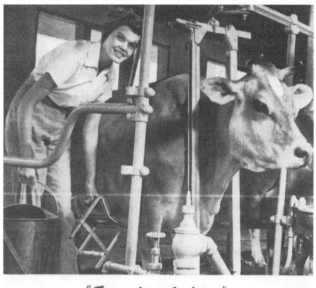

(a)   *"Front line fighter"*

(b)   *"White Ammunition"*

Figure 3.12 **Patriotic milk promotions in the U.S. during World War II** (Source: (a) *Journal of the American Dietetic Association* (September 1942). (b) *Journal of the American Dietetic Association* (July 1942)).

In addition, diverse arms of the military sought to provide at least the domestic troops and installations with a variety of fresh foods, including milk (Rifkind 2007). Increasing milk production and consumption took on a nationalist rhetoric during World War II, with milk promotions emphasizing milk's associations with strength and power (a theme we will return to in Chapter 5; see Figure 3.12). However, the end of the war brought renewed availability of prized foodstuffs and declining needs of the military, and milk prices fell as the supply exceeded falling demand. Even so, milk intake never regained its pre-World War II popularity.

Current annual milk intake is now at half the level it was at its peak in 1945 and about two-thirds of the 1909 level. However, Figure 3.11 also shows that while Americans are drinking less milk, they are eating much more cheese. Annual per capita cheese consumption rose by a factor of eight (4 to 33 lb; 1.8 to 15 kg) since the early 20th century. Looking more closely at the past 40 years (Figure 3.13), while total dairy product consumption is currently about 80% of what it was in 1970, milk consumption has declined more steeply and is now about two-thirds of its 1970 level. In contrast, cheese intake has almost tripled. Yogurt has had a more spectacular increase but the per capita level of intake is still quite small (8 lb [3.6 kg] per year). Thus, while Americans are no longer

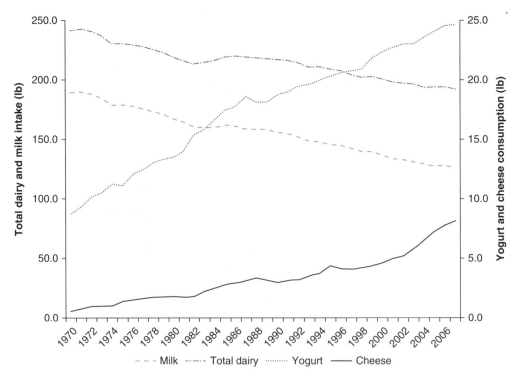

Figure 3.13 **Intake of total dairy, milk, yogurt, and cheese (lb), 1970–2007** (Source: Data derived from United States Department of Agriculture (2008)).

consuming milk in a glass, they are increasingly eating it on their plate, in a variety of cheese-laden dishes. Furthermore, while consumption of whole milk has been giving way to lower fat milks across all age groups, this trend is co-occurring with an increase in consumption of high-fat cheeses.

Americans are not alone in their newfound fondness for cheese; Europeans have increased their intake of cheese as well, now consuming ~20 kg (44 lb) of cheese per year, up from around 10 kg (22 lb) in 1977 (EuroFIR n.d.). Milk consumption has likewise declined in Europe, but remains above that of the U.S.

Decomposing the trend toward reduced milk intake indicates that the decline has been more precipitous among children (Figure 3.14; these are individual-level survey data in contrast to national per capita data shown in Figures 3.11 and 3.13). While the data in Figure 3.14 go up only to the mid-1990s, data from 2004 to 2005 indicate a continuing of that trend such that 6- to 11-year-olds consume only 257 grams (just over one cup) and 12- to 19-year-olds drink 178 grams (0.7 cups) and adults 20+ years drink 106 grams (0.4 cups) (National Center for Health Statistics 2005). Those who had been the featured target for milk promotion are now less likely than ever (at least since the late 19th century) to consume milk, although as Figure 3.14 shows, they still consume greater quantities of milk than do older age groups.

What factors are responsible for these changes? Certainly the proliferation of other drink types contributes, although the downward trend began long before

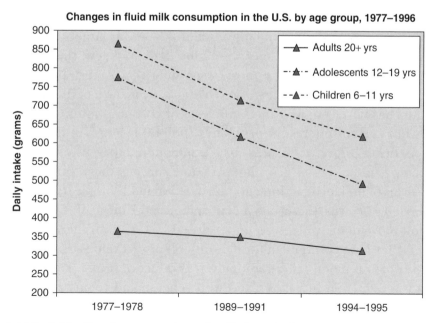

Figure 3.14 **Individual-level consumption of milk has declined most precipitously among children** (Source: Drawn from data in Enns et al. (1997, 2002, 2003)).

these appeared. Soft drinks and other "sugar-sweetened beverages" are now more widely available and contribute more calories to the diet than milk (Popkin 2010).[2] The sheer variety of drink options emphasizes to consumers that options other than milk exist. While milk is still ubiquitous among school feeding programs, schools have signed pouring rights contracts with soft drinks companies that allow the latter to place vending machines in schools in return for funding (Nestle 2002). Eating outside of the home is another phenomenon likely linked to declining milk intake, as consumers—including children—are able to choose what they like and are encouraged to "indulge." Milk must now compete for a "share of the throat" (den Hartog 2001).

These trends are mimicked to some extent in advertisements for milk. While the milk industry has not given up on children as a market for its products, there has been a substantial effort to increase consumption among adults. Undoubtedly this reflects recognition of the changing demography of the U.S. As fertility has declined and longevity increased, there are fewer children and more older Americans who represent an untapped market for milk. Recent advertisements and websites have been highlighting adults, although many are targeted at mothers, both for their own consumption and that of their families. As we will see in Chapter 5, these promotions highlight milk as the ideal food to fix a range of contemporary social and health problems.

Given that this chapter has documented the relative rarity of fresh, fluid milk consumption in the history of the U.S. and Europe, and the predominance of cultured milk products and cheese, these recent trends represent—in some ways—a return to these earlier patterns. Cheese, long used as a means of preserving milk, is consumed in larger quantities, and is largely attributable to the trend toward eating outside of the home. Italian cheese (i.e., mozzarella) consumption has increased the most, suggesting that Americans' affinity for pizza underlies this trend. What is also striking is the growth of the yogurt industry. Yogurt has never enjoyed the same level of consumption in the U.S. as in Europe, but current interest in probiotics (helpful bacteria in the colon) and their benefits to "digestive health" for a population apparently suffering from constipation underlies its growing presence among dairy products. While earlier forms of milk production and preservation made ample use of fermentation by such bacteria, these had been denigrated during the heyday of the sanitation movement (which was undergirded by the germ theory of disease). However, they have resurged, although in high-priced value-added dairy products. It is worth remembering too that a byproduct of fermentation or cheese making is a reduction in lactose, and while cheese and yogurt are not specifically marketed for this characteristic, it may facilitate their consumption among diverse minority groups with high rates of lactose impersistence. While cheese is consumed across the age spectrum, probiotics are marketed primarily to adults (and among them, women). Heavily sweetened yogurt is readily available and marketed to children.

Although nutritionists and the dairy industry alike decry the decline in milk consumption among Americans, it may be that widespread consumption of fresh milk was more of a historical interlude than an accurate representation of the "normal" American diet, spurred on by the novelty of easily available and safe milk, marketing of milk as a nutrient-rich food during a time when nutrients were just being identified, and support by the government through dietary recommendations and feeding programs. Surely fresh milk will not disappear from the American diet, and it remains a normative drink for children, but its predominance as a drink of the early and mid-20th century has come to an end.

4

MILK CONSUMPTION, CALCIUM, AND CHILD GROWTH

As we saw in the previous chapter, milk promotions in the U.S. historically have rested on claims that milk is a "special" food, one that enhances or is essential for proper child growth. These have come to include assertions about milk's importance to bone density, and to reducing the risk of osteoporosis and related fractures among older individuals, especially post-menopausal women. There has also been a recent attempt to link milk consumption to weight loss, although a lack of strong evidence has required a softening of this claim. There are many other statements out there concerning milk's impact on the risk of other diseases (cancers, cardiovascular disease, diabetes, hypertension, etc.), but to review them all would be well beyond the scope of this book. I will focus on the relationship between milk and child growth and "strong bones," as these messages have endured and seem most intuitive for many people. Since calcium is central to many of these, I will start with this mineral.

Calcium: Miracle Mineral?

As we saw in the previous chapter, the promotion of milk through the mineral, calcium, started in the 1930s, but really escalated in the late 20th century. It has become somewhat difficult to distinguish between milk and this mineral, as they are often equated with each other. For example, the "Milk Matters" campaign, which is sponsored by the National Institute for Child Health and Human Development, is ostensibly about milk, but in fact the real focus is on the importance of calcium to growing strong bones. [1] Milk is highlighted as the key way to increase calcium intake. Many of the "got milk?" ads likewise feature calcium, and it is this mineral that parents worry about if their child doesn't like to drink milk. In a recent conversation I had with a nutritionist about my research on milk and child growth, the nutritionist said, "Oh, you mean you're interested in calcium." Recall the earlier discussion of nutritionism: milk has come to be defined largely in terms of this mineral, although currently Vitamin D, which is added to milk, is receiving increasing attention.

It is worth keeping in mind that milk, like any other food, is a source of calories (from lactose, fat, and protein), although that varies depending on whether it has had some or all of the fat removed. Whole milk has 146 kcal/cup (245 g), while non-fat (skim) milk has 86 kcal/cup (United States Department of Agriculture 2010b). Milk is also rich in protein, with 8.2 g/cup. It should be noted that in the United States cows may be treated with synthetic bovine growth hormone (rbGH, also known as rbST), which may alter the protein profile of milk. In the U.S. milk from cows treated with rbGH does not have to be labeled as such, and so it is difficult to know how much of the milk people consume is from treated cows. In many other countries rbGH is not allowed, and its use has declined in the U.S. In addition, there are multiple other nutrients (see Figure 1.2) as well as bioactive compounds whose functions are not all well understood. In sum, milk is a very complex food, all the more so because of the conditions under which cows are maintained for milk production and the transformations it may undergo in processing, from fat removal to vitamin fortification.

The current justification for promoting milk rests almost exclusively on one nutrient in milk: calcium (United States Department of Health and Human Services and United States Department of Agriculture 2005), and the National Dairy Council proclaims that the United States, is facing a "calcium crisis" (National Dairy Council 2003b). This crisis, whether real or imagined, correlates with a decline in milk consumption and evidence that most U.S. citizens, milk drinkers included, consume less than the recommended daily allowance (RDA) for calcium (USDA Economic Research Service 2005). Not surprisingly, the solution to the calcium "crisis," at least according to the USDA and NDC, is to consume more milk rather than to consume other foods rich in calcium— fish, tofu, dark green leafy vegetables, or fortified non-dairy drinks.

An unresolved issue, however, is whether current calcium recommendations are reasonable, or whether they are too high (Anderson 2001; Hegsted 2001; National Institutes of Health 1994). If the recommended daily amount of calcium is too high, then more people will "fail" to meet the recommendation, and be counseled to consume more dairy products. There is no international consensus on calcium requirements. Even among milk-producing countries, the United States and Canada have much higher recommended amounts than the United Kingdom and the European Union for both children and adults (children: 500–800 vs. 350–550 mg, adolescents: 1300 vs. 800–1000 mg, adults: 800–1200 vs. 700 mg, respectively: Food and Agriculture Organization and World Health Organization 1998). Much of the world's population consumes well below the U.S. RDA for calcium without apparent detriment (Food and Agriculture Organization 2002; WHO/FAO 2004). The World Health Organization acknowledges that 500 mg is probably sufficient for adults but nonetheless recommends 1000 mg (WHO/FAO 2004).

Rationales for U.S. dietary requirements for children that include three servings of milk (or dairy products) per day are based on the fact that most American children do not consume the recommended amount of calcium. In addition, milk and other dairy products contribute more than 70% of the calcium intake in the United States, hence:

> Achieving that level of intake without dairy products requires careful attention to [a] selection of foods that naturally contain some calcium and others to which it is added. Unfortunately, calcium is found in significant amounts in relatively few foods. These foods are not consumed consistently in large amounts by most of the population and tend not to be popular with children.
>
> (Goldberg et al. 2002: 830)

According to this author, contemporary children are assumed to dislike the non-dairy foods that are rich in calcium, and as a result they should instead be encouraged to drink lots of milk. However, this assumption fails to question whether milk's predominance among available calcium-rich foods is necessarily ideal or whether children's preferences for milk should be encouraged to the exclusion of other calcium-rich non-dairy foods.

It is also worth noting as well that estimates of calcium intake among our Paleolithic hunter-gatherer ancestors are high relative to those in the contemporary United States (~1960 mg/day compared to 750 mg/day: Eaton et al. 1999). Children's diets that most closely resemble those of ancestral hunter-gatherers (i.e., rich in non-grain plant foods, legumes, nuts, or fish) can contain sufficient calcium, as well as other nutrients, and in the case of plant foods, phytochemicals, in the absence of milk. Of course in a context where such foods are not promoted or valued as foods for children, and where milk is widely available and relatively inexpensive, milk appears to be the most palatable alternative to meet the U.S. calcium requirements.

Does Milk make Children Grow?

The emphasis on milk as a source of calcium especially for growing children raises the question of whether milk consumption enhances child growth. Milk was the earliest food consumed by mammals, and among increasing numbers of human populations, cows' milk replaces breast milk or formula after infancy and is drunk during childhood (World Health Organization 2003). Given that the milk of a given mammalian species should contribute to the proper growth of nursing juveniles, the question arises as to whether continued consumption of milk, especially one derived from a species that grows rapidly and to a much larger size, likewise enhances growth throughout childhood. It seems quite plausible that milk's nutrients and bioactive compounds collectively should support the growth of older children, albeit possibly in an attenuated way compared to nursing infants. In other words, milk contains nutrients that are utilized in the process of growth, but there may be other components of milk that allow it to have "special" growth-promoting qualities that are not found in other foods.

Frequently used growth indices for children include weight for age, height for age, and weight for height (or BMI), among others. These measurements are plotted on a graph with age on the x-axis and the percentile for the growth index on the y-axis (or the z-score, which is the deviation from the median, which is set at 0: see Figure 4.1), and compared to accepted standards for child growth. Many countries have developed local standards for growth indices, but there are also international growth charts. The World Health Organization (WHO) recently completed a multi-country survey of growth among healthy children, and developed an international growth standard from that (de Onis et al. 2007b; WHO Multicentre Growth Reference Study Group 2006). This standard should replace the widely used U.S. standard from the National Center for Health Statistics (NCHS), although, for children over the age of 5, it uses NCHS data. The WHO standard for growth among children under the age of 5 indicates a surprising degree of uniformity in the growth patterns of healthy children from diverse populations.

One of the problems with the existing U.S. NCHS standard was that when used to assess infant growth, healthy, breastfed infants appeared to experience growth faltering compared to the standard. As shown in Figure 4.1, around the age of 2 months, breastfed infants experienced less rapid growth and by

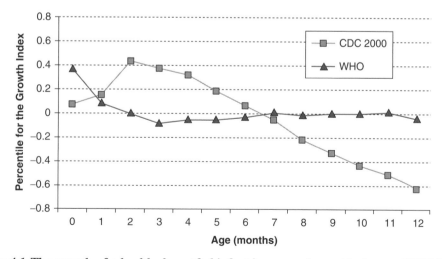

Figure 4.1 **The growth of a healthy breastfed infant in comparison with the new WHO infant growth standard and the previous NCHS/CDC standard** (Source: de Onis et al. (2007a)).

6 months their weight dipped below the 50th percentile (the median weight for age: de Onis et al. 2007a). Since their nutritional status was assessed in relation to the standard, many parents and pediatricians would become concerned that such infants were not consuming sufficient nutrients from breast milk, and a switch to or an addition of formula would be suggested. The WHO multi-country survey of growth revealed that the NCHS standard itself was creating this "problem." The U.S. children on whom it was based were largely fed cows' milk-based formula, as opposed to breast milk, and this was reflected in their more rapid growth. The new WHO growth standard which derived from that international study is based exclusively on healthy, breastfed babies. As predicted, the growth of a healthy, breastfed baby tends to be consistent with the new standard (Figure 4.1). Instead of the growth pattern of formula-fed babies serving as the standard, they are now seen as experiencing exceptionally rapid growth to excessive sizes.

Extrapolating from this example, the question arises as to whether children who drink milk beyond weaning likewise have accelerated growth compared to those who do not drink milk. The problem is that it is difficult to find a population whose children are healthy, with no major parasites, heavy infectious disease burden, or undernutrition due to chronic food scarcity, and who also do not drink milk these days (the globalization of milk consumption is discussed in Chapter 5). Furthermore, older children are eating a mixed diet in which milk is consumed along with many other foods, so it is tricky to ascertain the contribution of milk to their growth, which is likely to be small. Nonetheless, it is useful to ask the question, especially when coming from a cultural context in which it is assumed that children need to consume milk to facilitate their growth.

As we saw in the previous chapter, milk was not always a routine component of children's diets, even in countries where dairy products were an important aspect of diets in general. Fresh milk consumption became more widespread in the early twentieth century, and it is common to find assertions attributing the well-documented increases in average height during this time in the United States (and also European countries) to greater milk consumption. For example, Stuart Patton, an esteemed physiologist who specializes in lactation, writes:

> While many factors have contributed to this increase, it is obvious that calcium in the diet would be essential, and that products of the expanding American dairy industry would be the logical source of the calcium enabling this growth … . Of course calcium, while essential to increased bone growth and stature, is not the only contribution that milk would be making in this situation. High quality protein and growth-promoting B vitamins and Vitamin D from milk would be other contributing factors.
>
> (Patton 2004: 115)

Likewise, in Japan, Takahashi claimed that milk "Seems to be the most effective food for stimulating growth in height … . The short stature of Japanese in the past may be mainly caused by this low calcium diet" (1966: 125). As these quotes also indicate, calcium is considered to be the milk component responsible for milk's contributions to increased height.

There are strong historical correlations between increases in milk intake and height. In Europe, greater height among Northern Europeans has been attributed to their more extensive involvement with cattle raising and use of milk compared to Southern Europeans (Köpke 2008). According to this author, "autochthonous Germanic people in *Germania Magna,* beyond the borders of the *imperium Romanum,* were taller than in the core-land of the empire because they produced and consumed more milk and beef" (Köpke 2008: 139). Furthermore, the author claims that "cattle were certainly used for both milk and meat, but milk has a stronger influence on regional human nutrition." Recall from Chapter 3 that the widespread use of refrigeration to transport meat and milk in the late 19th and 20th century was also highlighted as a contributor to increases in height (Craig et al. 2004).

These studies have attempted to statistically tease out the specific role of milk as the most important factor to greater height, but the 20th century increase in height in particular occurred amidst a downward trend in childhood infectious diseases and improvements in nutrition due to overall increases in the food supply (Bogin 1998). Height increases were associated with overall improvements in health, and height is widely used as a measure of population health (Komlos 1994). Moreover, it is worth pointing out that milk consumption has been declining since the 1940s, and is now well below where it was at the beginning of the 20th century, yet height has not simultaneously declined.

Claims that milk enhances growth in height are also common in milk advertisements and promotions. These have existed throughout the 20th century, as the image in Figure 4.2 from 1947 shows. More contemporary images of professional basketball players are depicted with milk or statements such as "Hey everybody! Want to grow? About 15% of your height is added during your teen years and milk can help make the most of it. Milk rocks!" or "The calcium and Vitamin D in milk work together to help you build better bones and grow tall" (Milk Processor Education Program n.d.a; Milk Processor Education Program n.d.b). Surveys of university students (Wiley n.d.) have found that the majority of respondents highlighted bone growth as a positive outcome of milk consumption. More common are statements to the effect that milk helps you "grow strong bones," a conflation of milk's purported effects on both bone size and density. Interestingly, milk promotions emphasizing milk's contributions to growth in height have faded from view over the past five years, and have been replaced by milk's contributions to physical fitness and healthy weight.

What a difference from their daughters!

Figure 4.2 Advertisement from the National Dairy Products Association. Text reads: "There've been some changes, all right, in the past generation ... For one thing – Betsy Co-ed's grown a good bit taller since mother played center on the team! ... This doesn't mean we're raising a race of Amazons ... . But it does mean younger folk have better foods to 'grow on' than their parents ever did ... . Many of them are developments of National Dairy Laboratories ... that milk, nature's most nearly perfect food, offers virtually all the raw materials of modern nutritional research." (Source: *Journal of the American Dietetics Association* (1947)).

Studies of Milk Consumption and Growth in Height

What is the evidence for a link between milk consumption among children and growth in height? Given proclamations that milk has made children grow taller throughout the 20th century and popular assumptions that milk "makes children grow," one would think that there would be extensive evidence supporting this conclusion. In reality there are relatively few studies, and even fewer that are sufficiently well designed to allow us to conclude anything about milk's "special" effects on growth.

There are two types of published scientific studies that assess the relationship between milk and growth. One is an observational study, in which a large survey of individuals for whom there are data on milk consumption and height is analyzed to see if there is a positive correlation between milk and height or if variation in milk consumption explains some of the observed variation in height. These types of studies cannot demonstrate that milk *causes* any differences in height that might exist among people with high or low milk consumption levels or frequent vs. rare milk consumption, but rather they can only show that those with more/less milk consumption are, on average taller/shorter, and in large studies, other factors that influence height may be held constant (parents' height, overall calorie intake, etc.) to ascertain milk's contribution to variation in height. Milk intake is most often measured as the amount consumed in the past 24 hours or over the course of a few days, or as reported frequency of milk intake over a defined period of time (week, month, year).

The other type of study is a controlled supplementation trial. A sample of children who are similar in age and baseline dietary practices is divided into two groups. One group receives a milk supplement of a defined amount, usually on a daily basis (the "experimental" group), and the other receives nothing (the "control" group). Children are measured at the beginning of the study, their baseline diet is recorded, and then they are followed for a period of time (at least several months). At the end of the study children are measured again. The difference between their starting and ending height is calculated and the control and experimental groups are compared. If there are no other differences between the supplemented and control groups (e.g., the control group did not routinely drink more milk than the experimental group), any differences in height between the two groups should be attributable to the greater milk intake. Thus this type of study does allow us to say that milk is responsible for differences in growth, if any are found. However, if the control group received nothing, these studies do not allow us to draw any conclusions about *how* milk led to greater height. To assess this, you would need to have other comparative groups that received a calorie, protein, fat, calcium, or any other milk nutrient supplement, to establish whether it was any of these nutrients, or some other component of milk, that was responsible for the differences in height. In reality these types of studies are virtually non-existent.

One of the best studies to date was actually done in the 1920s, just after McCollum and others' work had demonstrated that rats grew larger on a milk-based diet. In this study, groups of British school children aged 6, 9, and 13 years were subdivided into groups provisioned with a glass of whole milk, skim milk, a biscuit equivalent to the calories in the skim milk, or nothing on a daily basis for seven months. The results shown in Figure 4.3 indicate that children receiving a daily milk supplement grew modestly more over a seven-month period than those who received a biscuit with the same amount of calories as the skim milk (Leighton and Clark 1929). Since calories were controlled for in this comparison, and the children receiving full-fat milk did not grow more than those consuming fat-free milk, it was concluded that there was something "special" about milk that contributed to growth in height above and beyond its caloric value.

Subsequent follow-up studies done in the United Kingdom in the 1970s and 1980s did not, however, confirm these results (Cook et al. 1979; Rona and Chinn 1989). Baker and colleagues (1980) found that giving 190 ml of free milk

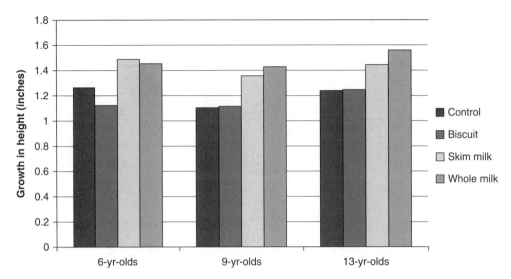

Figure 4.3 Study published in the Lancet in 1929 demonstrating that children receiving a milk supplement grew more in height over a period of seven months than those who received nothing or a biscuit of caloric value equal to the skimmed milk supplement (Source: Drawn from data published in Leighton and Clark (1929)).

to 7- to 8-year-old British schoolchildren classified as "disadvantaged" over almost two years produced a very slight increase in height. Children in the provisioned group grew 2.9 mm (0.11 inches) more than those in the control group, who received no food supplement.

There have been several more recent studies designed to test the effect of milk supplementation on changes in various aspects of bone biology, which also report on changes in height during the supplementation period. These are summarized in Table 4.1. None have demonstrated a positive, statistically significant effect of milk intake on growth in height. The studies ranged in duration from one to two years and included girls of European ancestry from the ages of 6 to 16. The long-term study by Bonjour et al. (1997, 2001) considered calcium in milk extract, not milk specifically, and found no effect on height. Importantly, these studies have all been done on relatively well-nourished populations with traditions of routine milk and dairy product consumption.

Research in less-well nourished and non-Western populations is likewise inconclusive. Lampl and colleagues (1978) found that school-age Bundi children in New Guinea supplemented with skim milk powder grew almost twice as much in height (~1.5 cm) over a period of eight months as did children who did not receive the supplement. The majority of children in the study had experienced extreme growth stunting, and the authors attributed the increased growth among the milk-supplemented groups to the addition of milk protein. However, milk powder is not a protein isolate, and the control group received

Table 4.1 **Intervention Studies of the Relationship between Milk and Growth and Height**

Authors	Study population	Age group	Sex	Sample size	Intervention	Duration	Growth differential Intervention -control mean	p-value
Cadogan et al. (1997)	Sheffield, UK "White"	Mean= 12.2 years	Females	82	1 pint whole or low fat milk per day	18 months	0.7 cm	NS
Bonjour et al. (1997)	Geneva, Switzerland "Caucasian"	6–9 years	Females	108	850 mg Ca from milk extract per day	48 weeks	0.4 cm	NS
– <880 mg baseline Ca intake							0.6 cm	NS
– >880 mg baseline Ca intake							0.0 cm	NS
Bonjour et al. (2001)								
1-yr follow up of above							0.7 cm	NS
3- to 5-yr follow up of above							1.4 cm	NS
Chan et al. (1995)	Utah, U.S. "White"	9–13 years	Females	48	dairy products up to 1200 mg per day	12 months	0.4 cm	NS
Merrilees et al. (2000)	New Zealand	15–16 years	Females	91	dairy products up to 1000 mg per day	24 months	–0.3 cm	NS

Note: NS = non-significant, p<0.05.

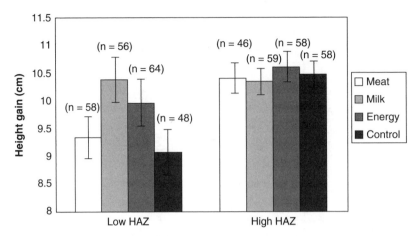

Figure 4.4 Well-controlled study from Kenya found no "special" effect of milk supplementation on growth in height of schoolchildren, even those with marginal nutritional status at baseline. High HAZ are relatively well-nourished (z-score>1.4). Low HAZ children are those with z-scores 1.4 below the median height at baseline (Source: Grillenberger et al.(2003: 91)).

no supplemental nutrients at all, so it could have easily been simply the addition of overall nutrients that allowed the children to grow taller. Among a large sample of 7-year-olds in Kenya, those who were given milk daily for two years did not grow significantly more than those who were given meat or fat (both with caloric value equivalent to the milk), or those who received no additional food (Grillenberger et al. 2003). As Figure 4.4 shows, only those children supplemented with energy or milk who had experienced growth stunting prior to the study (see left panel of Figure 4.4) grew significantly more (1.3 cm) than those in the control group. These studies suggest that children of marginal nutritional status may benefit from supplementation, but as the latter study shows, milk supplementation did not have greater positive effects on height than a basic energy supplement, and meat had no impact at all.

In China, a recent study indicated significantly positive effects of a school milk intervention on growth in height among adolescent girls over a two-year period (Du et al. 2004). Subjects were given 330 ml milk (fortified to contain 560 mg calcium) five days per week or no supplement. Baseline intake of milk was 113–135 ml/day in both the experimental and control groups. Girls who were given the fortified milk grew significantly more (0.7 cm; 0.27 inches) over two years. While these results are statistically significant, the absolute amount is modest, and since the control group received no alternative food, it is impossible to conclude anything about why the girls in the milk group grew taller.

While controlled supplementation studies are the most powerful test of a milk-growth relationship, a number of observational studies exist that support a link between milk and height. Based on their work among the Turkana pastoralists of Kenya, Michael Little and colleagues (Little et al. 1983; Little and Johnson 1987) proposed that the high protein intake of the Turkana contributed to their relatively tall stature, despite the fact that they consumed relatively few calories. The Turkanas' diet is based heavily on animal foods including dairy products (see Figure 4.5). Takahashi's global review of height in relation to ecological factors likewise found that populations with the greatest achieved heights were also those relying heavily on dairy products, although he was not able to control for overall nutrient intake (Takahashi 1971). In a different paper he noted that increases in height among schoolboys in Japan paralleled the increase in milk production in the country (Takahashi 1966), and in later work (1984) he reasoned that, by a process of elimination, the addition of milk through school lunch programs was responsible for increases in height among boys from the 1950s through the 1970s.

In a New Zealand sample of 250 children aged 3 to 10 of European descent, the 50 children who were classified as "milk avoiders" (defined as those who had avoided milk for at least four months at some point prior to study) were significantly shorter (3.0–4.1 cm) than 200 matched milk drinkers from the same town (Black et al. 2002). It should be noted that many of the "milk avoiders" suffered from allergies/asthma, and hence there are other variables (e.g., steroid treatment) that might have contributed to their shorter stature.

Milk's contributions to growth in height may vary by stage of growth. In an analysis of the large-scale National Health and Nutrition Examination Survey (NHANES) form the United States, Wiley (2005) demonstrated that milk intake among 5- to 11-year-olds was not related to height after controlling for calories consumed in the past 24 hours and birth weight. On the other hand, among

Figure 4.5 **A Turkana woman in Kenya milking her cow** (Source: Photo courtesy of J. Terrence McCabe. Used with permission).

preschool-age children, milk consumption was positively associated with height, after taking into account total energy intake, birth weight, sex and ethnicity (Wiley 2009). Among adolescents aged between 12 and 18, individuals reporting greater frequency of milk consumption were significantly taller than those with lower frequency of milk consumption. This effect was small with each increment of increased frequency (a scale of 0 = never drink milk to 4 = everyday) contributing about 3 mm to height (Wiley 2005). Similarly, a recent study by Catherine Berkey and colleagues (2009) followed U.S. girls from the ages of 9 to 11 through adolescence, tracking their self-reported milk intake and height. As shown in Figure 4.6, at ages 9 to 11 there were no differences in height among those who drank less than one serving of milk and those who drank more than three. However, those drinking more milk grew taller during early adolescence. Their growth spurt stopped at the same time as that of the girls who drank less milk, but the net effect was greater attained height among those who drank the most milk.

Milk's contributions to growth thus appear to be largely restricted to periods of particularly rapid growth and may not be as visible among populations with uniformly high levels of milk intake. The positive association between milk and

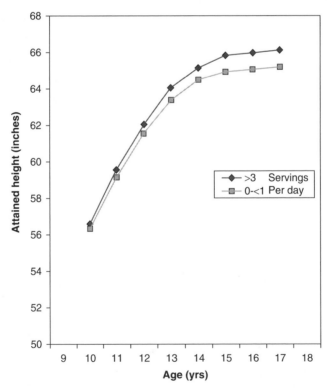

Figure 4.6 **Difference in height growth curves between white, non-Hispanic girls who drank >3 servings of milk per day and those who drank** (Source: Reproduced and adapted with permission from the American Association for Cancer Research: Berkey et al).

height among very young children may reflect a growth pattern attuned to milk consumption (i.e., evolutionarily children in this age group would still be consuming some breast milk) and that they (and adolescents) are growing rapidly compared to primary school-age children and can thus convert milk's components into linear growth. They may be physiologically more sensitive to milk's properties or they may simply show more variability in growth.

Interestingly, in the NHANES analyses, the total amount of dairy products children consumed had no relationship to height among any age group (Wiley 2009, 2010). The associations between dairy and height appear to be specific to milk, and not attributable to the components of other types of dairy products such as cheese, yogurt, or ice cream. This suggests that there may be something unique to milk that contributes to growth and that acts independently of caloric value, calcium, or protein content among young U.S. children. It is either not present, present in much smaller quantities, or deactivated in processed dairy products or during the digestive process.

What components of milk might be responsible for growth in height? Calcium is a major component of the inorganic matrix of bone, and hence is necessary in the process of bone growth, but most studies have found that calcium supplementation does not have positive effects on children's height (Bonjour et al. 1997; Cameron et al. 2004; Dibba et al. 2000; Gibbons et al. 2004; Lee et al. 1994, 1995; Winzenberg et al. 2007). In the NHANES study overall calcium was positively associated with height among preschool children, but children who drank the most milk were still taller than those who drank less after controlling for calcium intake. It is reasonable to conclude that calcium plays a role in the growth of height of young children, but it is not the sole or most important component, and it may act synergistically with other properties of milk.

Another component of milk that has attracted attention in the scientific community, although it is not highlighted in milk advertisements, is insulin-like growth factor I (IGF-I), which is part of the protein fraction of milk. Individuals who consume more milk have higher levels of circulating IGF-I in their bloodstream and supplementation with milk generates higher IGF-I levels (Cadogan et al. 1997; Hoppe et al. 2004; Zhu et al. 2004). Milk consumption, circulating IGF-I, and height have been found to be positively correlated among children of varying ages (Garnett et al. 1999; Hoppe et al. 2004; Rogers et al. 2006). It is not clear how intake of other dairy products affects circulating IGF-I. In a recent study of European women, milk and cheese, but not yogurt, were positively associated with IGF-I levels (Norat et al. 2007); a British study of adult men found that IGF-I was positively related to both total dairy and milk (Gunnell et al. 2003).

Since IGF-I is a protein, and should be digested like any other protein in the diet, it is not clear if it is the IGF-I in milk that increases IGF-I levels in those who drink more milk or if milk consumption stimulates one's body to produce more IGF-I (Holmes et al. 2002; Juskevich and Guyer 1990; Rich-Edwards et al. 2007). IGF-I is produced in bone cells and is the most abundant growth factor present in bone tissue; it increases the uptake of amino acids and contributes to collagen synthesis, it is involved in calcium and phosphate metabolism, and contributes to the growth and differentiation of bone cells and the formation of the bony matrix (Cameron 2002; Kelly et al. 2003). If milk intake stimulates production of IGF-I in bone, it may be a mechanism by which milk can contribute to skeletal growth.

Many concerns have been raised about whether the use of synthetic bovine growth hormone (rBGH or rBST) in cows to increase their milk yield would increase IGF-I concentrations in milk. Monsanto, the corporation that owns a patent for rBST, has funded research indicating that there is no difference in IGF-I concentrations in commercially available conventional milk, milk labeled as from cows not treated with rBST, or organic milk (which by definition cannot be from cows treated with rBST: Vicini et al. 2008). The Food and Drug Administration's conclusion is that milk from rBST-treated cows is safe for human consumption and that IGF-I levels are not elevated compared to either human milk or non-rBST-treated cows' milk, and IGF-I is digested and hence not directly absorbed (Juskevich and Guyer 1990). Not everyone is convinced that the studies are valid, and the controversy remains (Epstein 1996). There have been no studies comparing the growth of children consuming organic milk vs. those consuming milk from rBST cows.

In sum, observational evidence accumulated on the milk-height relationship does not point to a clear conclusion. At this point it appears that milk does not lead to greater height among primary schoolchildren, but it may do so among rapidly growing preschool-age children and adolescents. However, supplementation studies which have been done on primary school-age children and adolescents generally do not show an effect, either in Western milk-drinking populations or those without a tradition of childhood milk consumption. Few studies have compared milk to supplements of other nutrient-rich foods to ascertain whether milk has "special" growth-enhancing properties that are seen at different periods of growth and development. Most studies are done on children between the ages of 5 and 10 years, but given the relatively rapid growth of toddlers and adolescents, milk may be able to play a more important role in shaping their growth. IGF-I levels peak at around mid-puberty, which corresponds with a rapid rate of growth, and then decline throughout adulthood (Juul et al. 1994). Since milk consumption results in increased serum levels of IGF-I and IGF-I levels are already high during this time, milk may be able to have its most potent effects on growth during early adolescence. Thus, despite claims of a clear link, at present it is impossible to conclude with any certainty that milk has a clear positive effect on height or to ascertain which components of milk might play a role.

Milk and "Strong Bones"

The most frequently cited long-term benefits of childhood milk consumption are its positive effect on bone density and decreased risk of fracture (cf. Heaney 2000; National Institutes of Health 1994). However, studies of the relationship between milk consumption and bone density in childhood do not reveal consistent relationships—many find that milk supplementation transiently increases bone density (Bonjour et al. 1997, 2001; Cadogan et al. 1997; Chan et al. 1995; Du et al. 2004; Merrilees et al. 2000), or that milk consumption is positively correlated with bone density (Black et al. 2002), but other studies do not find this effect (cf. Heaney 2000; Lanou et al. 2005; Weinsier and Krumdieck 2000 for reviews). Furthermore, among studies finding that milk consumption in childhood had a positive effect of dairy foods on bone mass, the amount of variation explained by dairy food is extremely small (Weinsier and Krumdieck 2000).

The literature on milk consumption and bone health is large and contentious and beyond the scope of this chapter, but there are two points relevant to this discussion. First, osteoporosis, namely the loss of bone density that often occurs with aging, has come to be seen as a disease with its origins in childhood (Greer et al. 2006; Nicklas 2003). Because bone loss occurs from the total bone mass an individual has accumulated, the larger the bone mass achieved during growth, the lower (or later) risk of osteoporosis. Thus, understanding the determinants of child bone density is crucial to understanding this late-life disease. Several studies have shown that milk consumption in childhood is positively correlated with reduced risk of fractures in adulthood (Kalkwarf et al. 2003; Sandler et al. 1985; Teegarden et al. 1999), although other studies have failed to find evidence of this relationship (Feskanich et al. 1997; Lanau and Barnard 2008; Weinsier and Krumdieck 2000).

Second, it should also be recognized that other childhood behaviors can have a positive effect on bone density. Several studies have shown that childhood weight-bearing exercise has a strong positive impact on bone density (Anderson 2001; French et al. 2000; VandenBergh et al. 1995), independent of milk intake. Thus from an intervention standpoint, increasing weight-bearing physical activity among children can offset the later life risk of osteoporosis and bone fracture. Not all interventions need to involve increasing milk intake among children.

The "Calcium Paradox"

A paradox that some milk critics are quick to point to is the fact that the countries with the highest dairy product intake also have the highest rates of fracture and/or osteoporosis (Frassetto et al. 2000; Hegsted 2001), what WHO/FAO has dubbed the "calcium paradox" (WHO/FAO 2004). Furthermore, within the United States, bone fracture rates are highest among peoples of European descent, and lowest among those with African ancestry. This difference is often attributed to population variation in overall bone mass, as African-Americans tend to have greater bone mass than those of European descent (Aloia 2008; Bauer 1998; Ettinger et al. 1997). East Asians and Asian-Americans tend to have lower bone mass and calcium intakes (Walker et al. 2007), but also lower fracture rates than U.S. whites (Walker et al. 2008). Milk intake in childhood does not appear to be linked to adult bone density among African-American women, who report lower milk intakes than white women (Opotowsky and Bilezikian 2003).

There are many factors that influence the risk of osteoporosis or bone fracture via effects on bone density, including genetics, physical activity, exposure to UV light (as a marker of Vitamin D synthesis), weight and height, and intakes of Vitamin D, Vitamin A, calcium, fluoride, sodium, potassium, protein, alcohol, fruit and vegetables, among others. Thus, while calcium is important for bone density, it is not the sole or perhaps even the most important factor influencing it. Current research is pointing toward Vitamin D as a critical factor (cf. Cranney et al. 2008). The bone remodeling that contributes to bone resilience derives from complex interrelationships between calcium and Vitamin D (among other factors), which is likely to be more abundant given the higher UVB light exposure at lower latitudes, where milk is drunk less commonly (Durham 1991). Since milk is fortified with Vitamin D in the U.S., it is easy for milk promotions to continue to promote milk as important to bone health.

Fracture risk is also positively associated with consumption of animal protein across populations, and milk and animal protein intake are generally positively correlated (Frassetto et al. 2000). Furthermore, there are important interactions with physical activity, such that differences in subsistence activities contribute further to this population variation. Thus across populations lactase impersistence and a relative absence of dairy products in the diet do not necessarily contribute to increased risk of osteoporosis. What is also evident, however, is that as populations move to a more Western, industrialized lifestyle, which often includes dairy consumption, the risk of osteoporosis increases (Lau et al. 1990).

Milk and Weight

While strong bones have long predominated as a secure claim for milk, a variety of other claims are out there. Most surprising perhaps was the 2004 "got milk?" promotion featuring Dr. Phil (McGraw) that proclaimed "drinking milk can help you lose weight." New "three-a-day" and 24/24 (24 ounces of milk every 24 hours) slogans were in wide circulation on the internet, television, and print media in the United States, with slogans such as "Milk your diet; lose weight! Including 24 ounces of low-fat or fat-free milk every 24 hours in a reduced calorie diet provides calcium and protein to support healthy weight loss." With dire warnings about the obesity epidemic, this new portrayal of milk seemed poised to reinvigorate milk sales among perennially dieting but ever-fatter Americans. However, this promotion was pulled in 2007 when the U.S. Federal Trade Commission cited a lack of scientific support for the claim that milk consumption contributed to weight loss. While the explicit statements about milk and weight loss are gone, among the current "got milk?" celebrities are winners of "The Biggest Loser" weight loss competition shows, a current fad in reality television.

While the shaky scientific foundation for this claim plagued the promotion from the start, it is interesting to consider it in light of the "drink milk and grow" motif that had long been used. The transformation of milk as a food for growth to one that promotes weight loss was not necessarily contradictory. Milk may still be positively evaluated as a food associated with growth in height, which is highly valued in the United States (Fox 2005). Weight is a different matter, with ideal body weights well below the current norm. Furthermore, weight follows a class gradient in the United States, with current studies showing higher rates of obesity and overweight among the poor and ethnic minorities (Miech et al. 2006; Ogden et al. 2006). In contrast, height is positively correlated with income (Wells et al. 2008). Neither milk consumption nor its effect on height varies by economic status (Wiley 2005), but the ideal U.S. body shape and the ideal successful U.S. citizen is tall and lean, and milk could now provide both dimensions of that body type. What is also striking is that while growth in height can only accrue during childhood, weight is malleable across the lifespan, which greatly expands the market for milk. The current message at www.whymilk.com (sponsored by the National Milk Processor Promotion Board) is that milk "builds strong families," which emphasizes that milk is for all age groups.

In conclusion, while milk has been promoted for a variety of positive effects on body shape, size, or bone density, the evidence does not always point to a clear outcome. Most frequently milk is considered to have biological effects via its high calcium content, but as we have seen, the evidence for that is also not uncontroversial. While some relationships between milk and biology seem intuitive ("milk makes children grow"), in reality there exist relatively few studies that provide a test of whether milk has "special" effects on growth that are different from other foods.

5

GROWING CHILDREN
AROUND THE WORLD

The Globalization of Childhood
Milk Consumption

In his widely read book of essays *Good to Eat: Riddles of Food and Culture*, the late anthropologist Marvin Harris described population variation in lactose digestion and divided human societies into "lactophiles" and "lactophobes." He argued that not only were there population differences in the frequency of lactase persistence, but these tended to correlate with views about the "goodness" of milk. For Harris, the fondness for or disdain of milk was a profound delineator of cultural difference, with the lactophiles notably in the minority.

Over the past decades global patterns of milk production and consumption have shifted markedly, and this distinction between lactophobic and lactophilic populations is no longer so evident. Global trade in dairy products is at an all-time high, as is the overall volume of milk production and consumption, which has experienced tremendous growth in East and Southeast Asia and to a lesser extent in Latin America. With a rise in per capita incomes, many developing countries are said to be on the cusp of a "livestock revolution," manifesting as increased demand for milk and other animal-source foods (Delgado 2003). In addition, as noted in a 2007 *New York Times* article, in countries with emerging economies "[i]t turns out that, along with zippy cars and flat-panel TVs, milk is the mark of new money, a significant source of protein that factors into much of any affluent person's diet" (Arnold 2007).

The question arises as to why people in such countries have embraced *milk* as a key symbol of their economic success. How has this particular commodity, described as "wholesome" or "old-fashioned" by the U.S. dairy industry, become symbolically aligned with the hi-tech trappings of modernity? How has milk been able to transform from a marginal or disliked food to one positively associated with modernity and wealth? How has the dairy industry tried to refashion milk in the U.S. to stay relevant in the 21st century?

In this chapter I will first describe current trends in global trade, production, and consumption of milk, the ways that milk consumption is rendered normative (especially for children) by dietary policies, and then elaborate two case studies: India and China. Trends in these two countries will also be viewed in comparison with the U.S., where milk consumption is declining despite established beliefs about milk's "goodness," relatively high frequencies of lactase persistence, and a strong dairy industry with close governmental ties. Each country represents a different "package" of qualities related to global milk trends, and, given their size and growing affluence, India and China have attracted the attention of the global dairy industry. In China, milk was not part of the traditional diet and lactase persistence is very rare, yet currently milk consumption is experiencing the highest rate of increase of any nation. India is experiencing a similarly dramatic upswing in milk consumption, but dairy products have a long-standing central position in culinary and other aspects of culture there.

I argue that milk has been able to achieve its new status in India and China and attempts to maintain relevant in the U.S. by being positioned as a "special" food with properties that enable its consumers to achieve their goals for well-being, as defined in each context. As we saw in Chapter 4, one of milk's purported "special" attributes is that it enhances physical growth. Child growth becomes symbolic not only of the physical size of a country's citizenry, but also of national strength and power. In India and China (and early 20th-century U.S.), where child growth may be stunted from undernutrition and infectious disease, evidence of enhanced physical size provides confirmation of economic and social development and a thriving family. The fact that politically and economically powerful nations tend to have the highest levels of milk consumption and the tallest citizens creates a package of meanings that allows milk consumption to be an essential "mark of new money" in countries such as India and China. In the U.S., the milk–growth link has continued salience although it is no longer featured centrally in milk promotions, and this message is no longer sufficiently compelling to an aging population. Concerns about "excess growth" and rising rates of obesity and chronic disease have required a reworking of the milk message to address these "diseases of modernity" (Cordain et al. 2005).

Milk as a Globalizing Commodity

Steger's definition of globalization provides a framework for understanding how milk has become a global food. "Globalization refers to a multidimensional set of social processes that create, multiply, stretch, and intensify worldwide social interdependencies and exchanges while at the same time fostering in people a growing awareness of deepening connections between the local and the distant" (Steger 2003: 13). These include greater embeddedness in relationships that cross traditional political, economic, cultural, and geographic boundaries. The globalization of commodities is characterized by transformations in production, transportation, marketing, and ultimately consumption. These processes are encouraged by neo-liberal international economic policies that promote free trade of commodities and the production, processing, and marketing of products by multi- and transnational corporations(Atkins and Bowler 2001: 38–39). In turn new international divisions of labor arise, and capital moves easily across national boundaries to where new markets may be found (or fostered) and labor costs are lower. Food is a special kind of commodity, one that is intimately tied to notions of self, regional, or national identity, and biological well-being, and as such, its spread through global networks is likely to trigger particularly profound cultural changes.

There are several key dimensions of the globalization of cows' milk: trade, production, consumption, and government support of expanding dairy markets.

Trade

The volume of the world dairy trade increased by 25% in just five years from 1995 to 2000, with continued growth in the 21st century (U.S. Dairy Export Council 2002: 279). As Figure 5.1 shows, there is a primarily one-way flow from larger milk-producing countries such as those of the European Union, New Zealand, and Australia. The United States is also a player, but has been slow to seek out global markets for dairy products. These are all countries that have long-standing histories of milk production. Given that fresh milk is highly perishable, bulky, and heavy, it seems an unlikely candidate for long-distance trade, but technological innovations such as ultra-high temperature pasteurization

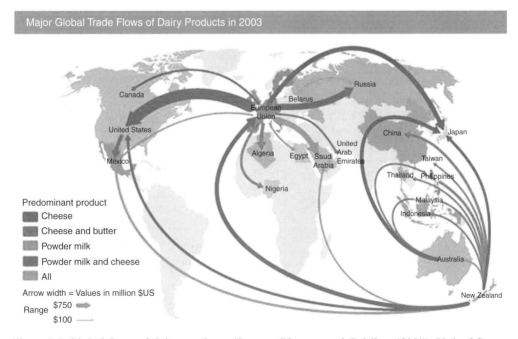

Figure 5.1 **Global flows of dairy products** (Source: Blayney and Gehlhar (2005). United States Department of Agriculture).

and freeze-drying have broadened its distribution. Only about 7% of total fluid milk production is traded internationally (Knips n.d.), and as Figure 5.1 indicates, milk products such as butter or cheese predominate in the global dairy trade. It is important to note that flows of dairy products are not solely dependent on the level of production and consumption in different countries, but are also affected by national and global economic policies regarding import tariffs. These can have the effect of limiting dairy imports as governments attempt to protect their own dairy producers (cf. Beghin 2006).

Production

The global production of milk has been increasing at 1 to 2% per year, with almost half of that growth now coming from developing countries (FAO 2008). As shown in Table 5.1, the European Union (27 countries) is by far the largest producer of fluid milk, followed by India, and the United States. Production expansion has been highest in developing countries, particularly in Asia, where production has more than doubled in the past 20 years, and continued growth is projected (Knips n.d.). From 2000 to 2009, China increased its fluid milk production over sixfold and surged well past New Zealand to become the world's fourth largest milk producer. Meanwhile India's total milk production more than doubled and surpassed that of the United States. Among traditional milk-producing countries, New Zealand expanded its production the most, followed by Australia and the United States. Mexico and Brazil both almost doubled their milk production.

Milk production and processing are increasingly controlled by transnational corporations. Three of the largest food corporations in the world (Nestlé SA,

Table 5.1 **Milk production and consumption in selected countries: 1990–2009**

Country	Total milk production (1000 metric tons)		Ratio of 2009 to 1990	Per capita kg fluid milk consumption		Ratio of 2009 to 1990
	1990	*2009*		*1990*	*2009*	
Brazil	14500	28795	1.99	60.9	55.5	0.91
Mexico	6456	11305	1.75	37.7	40.2	1.07
United States	67005	85820	1.28	105.1	91.2	0.87
Canada	7975	8200	1.03	100.8	92.7	0.92
EU-27*	–	137815			67.6	
New Zealand	7746	16601	2.14	124.7	79.3	0.64
Australia	6435	9670	1.50	102.4	110.0	1.07
Japan	8190	7900	0.96	41.0	33.8	0.82
China	4751	29625	6.24	2.5	8.9	3.55
India	53500	109200	2.04	30.6	39.5	1.29

* European Union, 27 countries. EU-27 did not report statistics in 1990.
Source: www.fas.usda.gov/psdonline/psdHome.aspx
Population estimates for per capita calculations from www.nationmaster.com

Kraft (which was split off from its parent corporation Altria in 2007), and Unilever) are heavily involved in the dairy industry, and have a presence in developing countries. The international dairy industry has also become consolidated among fewer larger firms over the past decade. Between 1998 and 2000 there were 415 mergers and acquisitions of dairy companies worldwide (Pritchard 2001). Nestlé is by far the largest dairy corporation, along with Dairy Farmers of America, Dean Foods, Danone, Kraft, and Parmalat.

Consumption

Growth in global production of fluid milk should be mirrored by rising levels of milk consumption. As Table 5.1 shows, annual per capita fluid milk consumption remains highest in countries where milk has been an important component of traditional diets. However, consumption trends in these countries have sustained either very modest gains or substantial losses. In Asia, milk consumption has declined from its post-World War II high in Japan, but surged in China and India. While China's per capita consumption is still very low, less than 10% of that in the United States or Canada, it more than tripled between 1990 and 2009, although it has not kept pace with production. Growth potential in the dairy market is clearly greatest in developing countries and those in which milk has not played a major role in traditional diets.

What forces are driving changes in milk consumption patterns? Dairy industry analysts cite demographic trends, notably the aging of the population, as well as increased competition with other beverages such as soda, bottled water, and juice as factors underlying declining milk consumption in Western countries, as we saw in Chapter 3 (Beghin 2006; Pritchard 2001). On the other hand, in developing countries such as those in Asia, younger populations, rising incomes and purchasing power, changing food preferences, and urbanization are seen as key to the boom in the dairy market (Beghin 2006; Delgado 2003; Pritchard 2001).

Milk in Dietary Policies

One mechanism by which milk is introduced and promoted in countries throughout the world is school milk programs. Growth and support of milk programs in developing countries is considered an important priority for the Food and Agriculture Organization (FAO) of the United Nations. In 2000, the FAO initiated a "World School Milk Day," which is celebrated worldwide once a year in late September. In addition to World School Milk Day, the FAO's Commodities and Trade Division coordinates and sponsors semi-annual international conferences on school milk programs at various international locations. Concerned with declining milk consumption in developed countries, dairy corporations are eager to support programs to increase milk consumption in developing countries and thereby enlarge the market for their products. The FAO's Commodities and Trade Division urged the dairy industry to "adopt an aggressive consumer-oriented marketing strategy to make milk and milk products attractive to school children" (Food and Agriculture Organization 1998). Providing school milk serves the twin goals of enhancing students' educational success through reduced hunger and better nutrition, and establishing a market for local and multinational dairy companies.

The FAO surveyed its 37 member countries to ascertain the extent to which schools have become a venue for the distribution and promotion of milk (Griffin 2004). Fifty-five percent made milk available in nursery schools or kindergartens, while 61% made it available in primary and secondary schools. Per capita weekly school consumption averaged between 0.06 liters to 1.25 liters. In most schools (67%) milk was either subsidized or provided free of charge. Dairy companies were often the main organizers of school milk programs, and in almost 75% of the responding countries, milk was promoted in schools by infrastructural support (refrigerators, dispensers), incentives, sponsorships, and educational resources.

Twenty-eight countries provided justification for the presence of milk in school programs. Over 86% noted milk as a rich source of calcium (although only 54% noted minerals in general), over 70% referred to milk as a source of vitamins, and 64% noted its role in a healthy diet. About half hailed milk's "good taste," but notably, the vast majority were countries in Europe or those with large European-derived populations.

Official dietary guidelines provide additional insight into the relative importance of milk or dairy in the diet. These are often generated by national governments or institutions with public health mandates, and may provide official sanction for feeding programs such as school milk initiatives. In cooperation with the FAO, the World Health Organization (WHO) recommends that countries develop guidelines that are appropriate to local economies, local health issues, cultural traditions, and those that are sustainable (World Health Organization 2004). Countries are urged to emphasize local foods, but the WHO and FAO acknowledge that countries are embedded in a global agricultural economy and that this inevitably affects food availability. A survey of current dietary guidelines from 12 countries (Australia, Canada, China, Germany, Korea, Mexico, the Philippines, Portugal, Puerto Rico, Sweden, the United Kingdom, and the United States) found that only the Philippines did not include a separate milk or dairy group in their guide (Painter et al. 2002). Milk is, however, included as an option in the major protein group.

Table 5.2 provides a larger sample of food-based dietary guidelines for countries in all regions of the world, and shows that the vast majority of countries with these include milk, although there is substantial variation in the

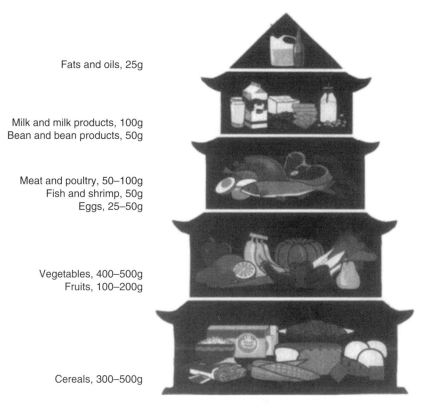

Fats and oils, 25g

Milk and milk products, 100g
Bean and bean products, 50g

Meat and poultry, 50–100g
Fish and shrimp, 50g
Eggs, 25–50g

Vegetables, 400–500g
Fruits, 100–200g

Cereals, 300–500g

Figure 5.2 **China's Food Pagoda.** (Source: Reprinted from *Journal of the American Dietetic Association Dietary Guidelines* and the Food Guide Pagoda. 100(8), Chinese Nutrition Society, pp. 886–888. Copyright (2000), with permission from Elsevier).

quantities recommended. Some of these guidelines were textual while others were pictorial, and the latter were much more likely to have an image of milk, cheese, or yogurt than the written guidelines were likely to list milk or dairy products. Milk was also likely to be listed as an option among other protein-rich foods or animal products rather than being given its own food group. This was particularly true in Asian and African countries, whereas in North America, Europe, Australia, and New Zealand, dairy products had their own dedicated food "group." In China's food guide pagoda (Figure 5.2) milk shares a category with legumes. In other countries such as Singapore, Nepal, Namibia, and South Africa, it is listed as an option among other animal products. Thus milk's protein content is the primary justification for its presence in international dietary guidelines, while it was calcium that was highlighted in school milk programs.

The food-based dietary guidelines listed here include only those targeted at the general population of adults and older children. However, the FAO survey indicates that 55% of responding countries reported that they had national recommendations for children to receive milk each day (Griffin 2004).

Table 5.2 **Country-specific Food-based Dietary Guidelines**

Country	Milk or dairy in Dietary Guidelines	Servings per day	Justification
Canada	Yes	2–4	protein, calcium
United States	Yes	3 cups/day	calcium, protein
Brazil	Yes	3 glasses	
Mexico	Not mentioned specifically		
Guatemala	Yes	1 at least 2 × per week	
Chile	Yes		
Argentina	Yes	1 serving	
Cuba	Yes		
Dominican Republic	Yes	3 × per week	
Australia	Yes	2–4	protein, calcium
New Zealand	Yes	Not specified	
Finland	Yes	~25% of diet	calcium
United Kingdom	Yes	"moderate amount"	calcium, protein, Vitamins B-12, A, D
European Union (member countries)	Yes—vast majority	200–750 ml	
Singapore	Mentioned in animal foods group	None required	protein, B-vitamins, iron, zinc
Thailand	Yes	1–2 glasses	
Philippines	Yes—option in animal foods group		calcium
Japan	Yes	2 servings	
Indonesia	Not mentioned		
Malaysia	Not mentioned		
India	Yes—option with grains and legumes	None required	
Nepal	Yes—option in animal foods group		
Bangladesh	Not mentioned		
Namibia	Yes—option in animal foods group	None required	
South Africa	Yes—option in animal foods group	2 cups	calcium

Source: ftp://ftp.fao.org/es/esn/nutrition/dietary_guidelines

These ranged from 0.2 liters (Japan, Kenya, Portugal) to around one liter (Canada, Saudi Arabia).

India and China: Promoting Milk through Growth

India and China represent two large countries with growing economies and increasing standards of living, whose per capita levels of milk production and consumption have increased greatly in the recent past (see Figure 5.3), in contrast to the U.S., which has had relatively steady milk production and declining milk consumption. A comparison of contemporary practices surrounding milk

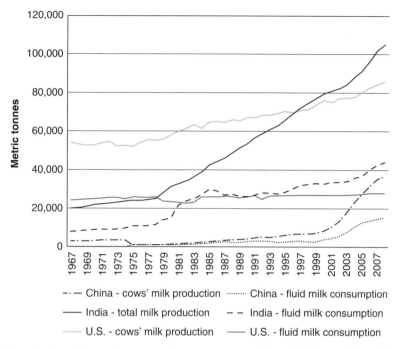

Figure 5.3 **Trends in milk production and consumption: China, India, and the U.S.** (Source: Drawn from data available from United States Department of Agriculture (2008)).

in these three countries, each of which has a distinct history with milk, provides insights into the local meanings of milk and how these intersect with globalizing processes that attempt to impose uniformity on views of milk. Milk production is a rapidly growing part of India and China's agricultural economies. India now produces over 20% more milk than the U.S., and China's production, which was negligible until 2000, is now around 35% of U.S. levels (see Table 5.1). China's citizens now consume an average of 9 liters per year (17 times what it was in 1970), while in India, per capita consumption has increased to 39 liters per year, 2.4 times what it was in 1970 (United States Department of Agriculture 2008). Both are still modest in comparison to the 90-plus liters per year drunk in the U.S. and suggest the tremendous potential growth of the market for milk in the largest of the world's countries. Meanwhile, in the U.S. where milk was an icon of modernity in the early 20th century (DuPuis 2002), "America's drink" struggles to maintain a place in contemporary diets. U.S per capita intake of fluid milk is about 76% of what it was in 1970.

In both India and China, milk had indigenous meanings that would seem to pose challenges to widespread adoption of milk as a marker of modernity. In China, milk was long reviled as a food of the "barbarian" Mongols, and ethnographers have commented on the absence of milch animals and widespread abhorrence of milk and its products there (cf. Harris 1985). In addition, most Chinese are lactase impersistent. India has low rates of lactase impersistence especially in the north and west, but notably higher rates in the south and east (Tandon et al. 1981). Milk products have long been an integral part of local foodways, and there is strong political, economic, and cultural support for milk and cows. Thus milk consumption would, at first glance, seem to be an odd part of a package that included sophisticated technological consumer durables.

The ways in which milk is promoted provide insights into the local meanings of milk as well as those that various institutions (government, industry, public health) would like to instill. In all three countries milk is promoted by the dairy industry, governments, and health officials, dieticians, and nutritionists. Government, NGO, and corporate-subsidized school milk feeding programs have existed since the 1920s in the U.S.; China established one in 1999 (Chen 2003b), and local school milk initiatives exist in many places in India. Tetra Pak, the manufacturer of aseptic packaging materials for milk, has long supported school milk programs in both China and India. Milk's privileged position in such programs is justified by its presence in governmental dietary advice. Milk is now listed in China's food guide pagoda (Figure 5.2) with these instructions: "Eat milk and legumes, and their products every day" (Ge et al. 2007), and is recommended in the Dietary Guidelines for Indians (National Institute of Nutrition 2006); India uses the USDA Food Pyramid in its promotions (see: www.foodpyramidindia.org).

Consistent with the results of the FAO survey and the U.S. Dietary Guidelines, the health benefits of milk and dairy products are most often attributed to their high calcium and protein content in these three countries. The Dietary Guidelines for Indians state: "Milk that provides good quality proteins and Calcium must be an essential item of the diet, particularly for infants, children and women" (National Institute of Nutrition 2006). Milk's inclusion in China's food pagoda is justified as "Products made with milk provide protein, vitamins and minerals, especially calcium" (Ge et al. 2007). The homogeneity of these messages reflects the global spread of scientific "nutritionism." This concept was introduced in Chapter 1 and refers to the privileging of nutrients as descriptors and markers of food value (Pollan 2008). The globalization of milk has been accompanied by nutritionist messages that aim to increase the desirability, and hence consumption, of milk by highlighting its valuable nutrients.

A related claim is that milk consumption is likely to generate increases in child growth. Here nutritionist views of milk emphasize calcium and/or protein as the important constituents of milk contributing to this outcome. The message that "milk makes children grow" appears to be particularly salient in global promotions of milk, and as noted in Chapter 4, milk "makes sense" as a food that would promote growth. Two aspects of growth are emphasized: growth in size (height) and strength ("strong bones"). These are different aspects of bone biology, but both can serve as powerful symbols for individual and national growth and strength. In China and India, claims about milk and child growth are widely articulated, but the meanings attached to these claims reflect local histories and contemporary goals for personal and national well-being. In the following sections, I consider the ways in which the milk–growth link is elaborated and made use of for particular purposes in China and India, and how these have emerged out of local historical conditions.

China

Cows' milk historically played a limited role in the Chinese diet. Most Chinese did not encounter milk after weaning, and the mutation conferring continued production of lactase had no opportunity to spread (Huang 2002). The absence of milking traditions appears to have co-occurred with an aversion to milk. The anthropologist Robert Lowie marveled at the "astonishing fact that eastern Asiatics, such as the Chinese, Japanese, Koreans, and Indo-Chinese have an inveterate aversion to the use of milk" (Harris 1985: 130). This quote is instructive for two reasons. First, Lowie found it "astonishing" that East Asians were averse to milk, indicating a widely held normative view by peoples of European descent (even among anthropologists in the mid-20th century!). Second, the Chinese would have fit squarely into Marvin Harris's "lactophobic" category. Milking was practiced by nomadic peoples in the north and west but after the Mongol dynasty ended in the 14th century, use of milk products dwindled (Huang 2002). Eugene Anderson suggests that this outcome was part of a nationalist agenda of the Ming to denigrate foods associated with the foreign "barbarian" Mongols (Anderson 1987).

Against this historical background milk has recently emerged, rather ironically, as a nationalist symbol. Chinese Premier Wen Jiabao was quoted in 2006 as saying, "I have a dream to provide every Chinese, especially children, sufficient milk each day" (British Broadcasting Corporation 2007). The Chinese have responded to government and dairy industry initiatives and are now said to have a "thirst for dairy" (Baumes 2004). The Chinese dairy industry has transformed from a miniscule enterprise to one that comprises 10% of the Chinese agricultural economy (Fuller et al. 2004), which is comparable to the United States where it accounts for about 11% (United States Department of Commerce 2002). The market for milk is focused in urban areas, where intake is more than 13 times greater than in rural areas (Baumes 2004; Lu 2009), and more common among higher income groups (Fuller et al. 2006). A survey of dairy purchases in households in three large Chinese cities found that over 90% purchased fluid milk, with an average purchase of just over one liter of fluid milk per week (Fuller et al. 2004). Dairy product sales in China have been growing at the rate of 25% per year, and the dairy industry is worth close to one billion U.S. dollars (Dairy Industries International 2006). Moreover, the industry is poised for continued expansion. As noted in a recent article in *China Today*, " It is estimated that only 300 million of the 1.3 billion people in China regularly drink milk, leaving a vast virgin market to be explored by dairy companies" (Lu 2009).

Rhetoric in milk promotions highlight milk's benefits to population health and includes the use of patriotic slogans such as "One cup of milk can strengthen a nation" (Chen 2003a). As Jun Jing notes, "the Chinese government has tried to educate the public to heed the officially identified need to improve the country's 'population quality' *(renkou sushi)*" (2000: 12). Education about proper child nutrition and promotion of milk is a core component of that mandate. Famous Chinese athletes such as Tian Liang, the Olympian diver, Liu Xiang, the Olympian hurdler, and triumphant Chinese astronauts fresh from their first successful spacewalk in 2008 are featured in milk promotions (see Figure 5.4). The use of athletes and those associated with national accomplishments on the world stage to advertise milk underscores milk's relationship to strength, size, and resilience, and serves nationalist, but also family and individual goals for strength and health. The vice-director of the Ministry of Agriculture only half-jokingly attributed China's soccer players' lack of international success to the fact that "they don't drink enough milk," and a spokeswoman for Shanghai

Figure 5.4 **Billboard featuring Liu Xiang, the Olympian hurdler, promoting China-produced milk** (Source: Getty Images. Used with permission).

Bright contended that consuming more milk will lead to faster growth rates among China's citizens and help make its men taller (Chen 2003b).

China's most internationally visible athlete, 7-foot-5-inch-tall National Basketball Association (NBA) player Yao Ming, was featured as the Western Conference "got milk? Rookie of the Month" in 2001 sporting a milk mustache. In January 2007, China's Mengnui Dairy took this link one step further by entering into a partnership with the NBA, and acquiring the rights to promote their products in association with the NBA's activities in China and other countries.

Successful internationally known athletes are a particularly effective marketing tool, especially for children and young adults, and further enhance milk's image as a "Western" food. As Brownell notes, "Dairy products are not a typical part of the Chinese diet. The emphasis on them for athletes is a result of an awareness that they form a large part of the Western diet, the assumption being that they explain the greater size and musculature of Western athletes" (Brownell 2005: 254). Milk thus offers a kind of message of hope in a bottle (or carton) for China's future as a robust, strong, healthy population and society.

Promoting milk by suggesting that it will enhance growth indexes local concerns about perceived Chinese "size deficits." Various Chinese officials articulate the view that the Chinese can "catch up" in size to Western populations by drinking more milk. For comparative purposes, average heights for men and women in China are 166–170 cm and 157–159 cm, respectively (Yang et al. 2005). In the U.S., the average height is 177 cm for men and 162 cm for women (McDowell et al. 2008), which results in a 7–11 cm difference for men and a 3–5 cm difference for women. Hopes for such an increase in physical growth among a new generation of Chinese children who drink milk are fueled by such marketing and bolstered by wide dissemination of the results of a school milk intervention study from Beijing (described in Chapter 4: Du et al. 2004) which indicated positive effects of a school milk intervention on growth in height among Chinese girls aged 10–12 with low baseline dairy and calcium intakes. Those girls who were given milk five days a week grew modestly more (0.7 cm) than a matched group given no supplementary food over two years.

India

Much like in China, Indian milk promotions contain references to height and growth, and blend these with nationalist goals. For example, a series of ads for the widely distributed "Mother Dairy" brand milk are quite explicit (Figure 5.5), featuring children in oversized professional clothes and a bottle of milk, and the bold slogan: "The country needs you. Grow Faster."

The association between milk and body size has long had political implications in India, which are best understood in relation to the deep history of milking both local zebu cattle (*Bos indicus*) and water buffalo (*Bubalus bubalis*).

Figure 5.5 **Mother Dairy brand milk advertisement from India** (Source: http://www. motherdairy.com/campaign.asp).

Cows were central to the subsistence and culture of the Indo-Aryan pastoralists who settled in the north of the South Asian subcontinent, and Vedic literature describes the high esteem in which they were held (Batra 1986). Cows were symbolically linked to the earth and mothers, both with life-giving properties (Korom 2000). The sacredness of cows, the laws protecting them, and religious rituals involving milk or *ghee* (clarified butter) all came to be core tenets of Hinduism. Hindu scriptures identify the cow as the mother of all civilization, whose milk nurtures the population. As such the cow became a potent symbol of "Mother India."

In the politics of late colonial India "the concept of 'mother cow' could be twinned with depictions of 'Mother India' and the life-giving, pure qualities of cow's milk could be associated in the minds of audiences with the purity and strength of the nation" (Gould 2004: 78). The cow and its protection became important metaphors for the colonized state and the strength of its citizens, especially its men. As Charu Gupta notes, "The cow was now [in the 1920s] more directly linked with building a strong nation, a nation of Hindu men who had grown weak and poor from lack of milk and ghee. For a body of healthy sons, cows became essential … . Like a mother, she could feed her sons with milk, making them stronger" (Gupta 2001: 4296). The need for the protection and improvement of cows to produce milk would correct "the poor physique of many of the population" (Home Poll 1922, quoted in Gupta 2001). Thus links between cows' milk, national strength, and physical growth have been articulated throughout the 20th century and remain potent metaphors. Notably, despite their economic importance, water buffalo and their milk never achieved similar recognition (Hoffpauir 1982; Mahias 1988).

A connection between milk and size is explicit in India, but talk about "catching up in size" does not prevail in milk rhetoric, despite average height in India being substantially shorter than in either China or the U.S. Average heights for men and women are ~155 cm and ~152 cm, respectively in India (Deaton 2008), about 12 cm shorter than men and 10 cm shorter than women in the U.S. (McDowell et al. 2008). Milk is not a novel food in the Indian diet and has long been valued as a food that is central to Indian dietary tradition and thus claims that milk is responsible for greater height in European populations are unlikely to be compelling. That said, greater height is a highly desirable trait for males and females, particularly among the upwardly mobile middle class. A brief survey of recent marriage ads from Pune (a city of 4 million in the state of Maharastra, and one of the major business centers of India) revealed that virtually all—whether it was grooms seeking brides or vice versa—indicated the individual's height, and when the desired height of the partner was indicated, it was always for "tall" regardless of sex. Height is a measure of the success of individuals and their families, and especially reflects on the quality of a mother's care (R. Parameswaran, personal communication).

More often than growth per se, strength—both physical and mental—is a common theme in Indian milk promotions. Even the ubiquitous "milk biscuits," which contain little milk, are said to be strengthening. For example, Milk Shakti (shakti means power) biscuits are advertised by M.S. Dhoni, the famous Indian cricket player: "From boy-next-door to Super Boy, no that's not the plot for the next Hollywood blockbuster it's the effect of Milk Shakti."[1]

On the other hand, lack of access to milk due to price or availability to the Indian populace provides a justification for the dairy industry's expansion, and if the growth and strength of its citizens is enhanced as a result, it will be further evidence of India's political and economic ascendancy in the post-colonial period. In 1970 the National Dairy Development Board (NDDB) initiated Operation Flood, which aimed to generate a "flood" of milk to supply Indian citizens and increase rural incomes by allowing small-scale producers to get a fair price for their milk by joining cooperatives. Import restrictions on dairy products were lifted in the 1990s, and there was increased private and multinational investment in the dairy industry. At present, dairy makes up about 17% of the Indian agricultural economy (Delgado and Narrod 2002), and increasing milk production is a main goal for Indian agricultural development. Milk consumption has been increasing in concert with a rise in incomes and purchasing power, particularly among a growing urban middle-class populace (Ali 2007). As in China, national data indicate that milk consumption is about 20% greater in urban than rural areas, and positively associated with socioeconomic status (Shetty 2002; Vijayapushpam et al. 2003).

Marketing Milk as Modernity

Beyond its links with growth, whether real or imagined, milk has other associations with modernity. For one, in all three countries, increased milk consumption has occurred amid economic expansion and urbanization, with urbanites the primary market for milk products, rather than in the bucolic, farm-based subsistence system that is ubiquitous as a marketing image. With ongoing rural–urban migration and the growth of cities across the globe, urbanites symbolize modern humanity. However, to become a viable product for urban markets, milk is subjected to a variety of technological transformations. While maintaining its "natural" superiority as a food, which requires at least some recognition of it as a bovine secretion, milk may be marketed as part and parcel of citizenship in the 21st century when it undergoes such alterations. Scientific interventions in milk production and processing are welcomed by a public concerned about milk's safety (and the safety of the nation and its economic progress), just as they were in the early 20th century U.S. (DuPuis 2002). Milk adulteration scandals in China have been widely publicized, [2] and people I have talked with in India acknowledge that their milk is more likely than not to be diluted or adulterated as purveyors try to maximize their profits.

Milk's transformations designed to meet consumer needs include pasteurization, cream separation (which enabled the production of skim milk and other fat-reduced forms, allowing milk to fit into a "healthful" low-fat diet), fortification (with Vitamin A, if fat has been removed, and Vitamin D), and homogenization, which causes the fat molecules to break up and become more evenly distributed in fluid milk, creating a more uniform appearance and richer taste, and renders milk more convenient to drink straight out of the carton ("without the inconvenient cream layer," as one Indian milk promotion asserts). Milk can be flavored to make it more palatable, and flavored milks (chocolate or fruit) are marketed heavily to children. For those who are lactase impersistent, lactose can be removed. Somewhat surprisingly, I have seen no Indian or Chinese promotions of such milk, although as we will see below, other lactose-reduced dairy products are easily available. In sum, the milk found in global supermarkets can hardly be considered a "natural" food; it has long been subject to technological modification to "fit" whatever health problems or target demographic is seen as most salient or lucrative at a given time.

Third, milk is linked to modernity via its associations with Western culture (and biology), economic progress, and wealth in China. In her study of children's diets in the Xi'an Muslim district, Maris Gillette remarked that,

> butter and milk in particular were regarded as typical of Western food. Even in snacks produced locally, the presence of dairy ingredients lent them an aura of foreign-ness and luxury. For *Hui* in their thirties and older, dairy products symbolized a high standard of living. Mingxin remembered craving milk as a child, when little was available and his family was too poor to buy it … [since] 1978, China's cows have been "science modernized" (*kexue xiandaihua*) and so dairy products were both readily available and affordable.
>
> (Gillette 2005: 111)

Indeed, it is the "Western" aspects of milk that are most salient to increasing demand for milk in China. As Gillette notes further with regard to Western factory-produced foods, "through these foods parents hoped to introduce their children to things foreign and equip them to live in an industrialized, technologically advanced, cosmopolitan world" (Gillette 2005: 117). That milk might enhance their growth (both physical and intellectual), allowing them to "catch up" in size to Westerners, further consolidates milk's image as a "modern" food and its consumption the means to the modern/Western life.

Milk's equivalence to modernity or Western lifestyles is more ambiguous in India in part due to long-standing cultural valuation of cows and milk. Milk has a variety of traditional symbols that could be used for its promotion, especially the cow. However, most milk in India derives from water buffalo, and, unless otherwise specified, all commercially available milk may be assumed to be a blend from both animals. This is not widely broadcast; instead happy cows predominate in milk advertisements, although it is important to note that they are not the indigenous beloved zebu cattle that serve as potent national symbols, but rather are the bigger foreign breeds such as the distinctive black and white Holstein well known in the U.S. and Europe.

Many Indian commercial dairies offer a separate cows' milk, but without any religious references. Mother Dairy notes only that its cows' milk is "easily digestible," which references Ayurvedic understandings of milk. Meera Premium note that their cows' milk is "processed in [*sic*] a very hygienic conditions in our production unit and contains calcium. It is supposed to be a good supplement for growing children and is considered very essential in the overall development of the health of growing children." Nandini likewise has a separate cows' milk, marketing it as "the good life" with an image of a Holstein cow. Milk is often advertised as "pure," which serves a double purpose of emphasizing its safety and referencing milk's religious significance. However, since "pure" is used for all milk products rather than just cows' milk, it is more likely simply a guarantee against adulteration.

Still, the brand "Mother Dairy" conjures up both the cow-as-mother and the nationalist "Mother India" motifs. Likewise Nandini is the name of a wish-granting cow, and also means "enjoyment" in Sanskrit. Text at the Nandini brand milk website merges the modern and traditional:

> Milk is nature's ideal food for infants and growing children in our country, except in rare cases of lactose intolerance. The important place milk occupies in our diet has been recognized since Vedic times, and all modern research has only supported and reinforced this view. In fact, milk is now considered not only desirable but essential from the time the child is born … . till he or she is 12 years of age.
>
> (http://www.kmfnandini.coop/html/knowyourmilk.htm)

While it may seem surprising that the Indian dairy industry does not make greater use of traditional motifs, the fact that water buffalo milk production predominates due to greater productivity is one obstacle. In addition, India has a sizable Muslim population (~140 million; ~14% of the total population) and strong nationalist Hindu political movements. The dairy industry must be careful not to incite communal antipathy by reference to religious symbols in its marketing. One solution is to divorce milk from its traditional roots and highlight nutritional science perspectives that emphasize milk's nutrients, while playing on some familiar—but not divisive—and/or unifying nationalist themes. By and large, the meanings of milk portrayed in milk promotions are squarely in the Western nutritionist mold with myriad references to milk as a rich source of protein and calcium and essential to building larger, stronger bodies.

U.S. dairy processors continue to struggle to find a "modern" image of milk. Like their Indian counterparts, they have chosen to blend the traditional with the modern. But what constitutes "modernity" in the U.S. is decidedly different from its positive valuation in China and India, with their burgeoning economies and rapidly rising standards of living. Milk promotions in the U.S. offer milk as a salve for the current *problems* of modernity—primarily the chronic diseases that threaten to reverse the gains in life expectancy that had typified the 20th century and a sagging economy conditioned to similar expansion. In addition, children, who serve as symbols of the future, are notably absent from the vast majority of milk advertisements (Wiley 2008). As children make up a smaller percentage of the U.S. population, and with ideas about milk and child growth and strong bones being well entrenched, the dairy industry is focusing its efforts on adults. The aging of the U.S. population and its attendant medical problems is reflected in the current image of milk, a true inversion of milk's evolutionary role as sustenance for the young.

The 2009 "Milk is nature's wellness drink!" promotion provides an illustration:

Milk may be nature's wellness drink that has been around for centuries, but the benefits are far from old-fashioned. It turns out that all milk drinkers have a leg up on other folks. For starters, studies suggest that people who regularly drink milk tend to have healthier diets that are richer in nutrients. They're also more likely to be at a healthy weight. Milk drinkers may also have an edge when it comes to cardiovascular health, including blood pressure. Plus, no other beverage offers the same array of bone-building nutrients, including calcium, vitamin D, protein and phosphorus, and has hundreds of studies supporting its link to strong bones. And there's also emerging evidence that the nutrients in milk may play a role in reducing the risk of cardiovascular disease, certain cancers, type 2 diabetes and metabolic syndrome.

(Milk Processor Education Program 2009)[3]

Middle-aged women (all of the milk-mustached celebrities used in association with this ad were adult women) are led to believe that milk provides a trusted source of goodness in troubling times. As Wendy Bazilian, a dietician writing for MilkPEP, notes:

When you think about all the ways you pursue wellness, drinking milk each day is one small step in the right direction. It's a simple thing you can do for yourself that can provide a powerful payback. With all the new age beverages on the market, it's hard to know what's best for your family, especially in these tough economic times. One you can feel good about is milk—it offers more nutrition for your dollar than virtually any other beverage you can buy.

(Milk Processor Education Program 2009)

Aside from the fact that "wellness" is sufficiently amorphous to preclude scientific evaluation, the promotion attempts a melding of traditional and contemporary themes. Milk has been around for centuries, unlike those "new age beverages" (it is notable that this statement was found alongside lithe women doing yoga, referencing both "new age" practices and ancient Indian traditions). Milk is also a panacea in "these tough economic times," providing a "powerful payback" of "more nutrition for your dollar." At the same time milk consumers will be sure to not suffer from the maladies that stem from the "modern" lifestyle. Thus milk, which served as an icon of modernity and technological advance and accompanied U.S. citizens into 20th- century affluence, unprecedented size, and longevity is now supposed to alleviate the problems attendant to these via its age-old value. But Americans appear to be unconvinced by this repackaging and milk consumption continues its downward trajectory. The early 20th-century proponents of milk for child growth may have unwittingly created an enduring meaning for milk which can spread across the globe but cannot find contemporary relevance among its local constituency.

Lactose Intolerance and Local Transformations of Milk

How has population variation in lactase persistence intersected with the global spread of milk? Consumption of milk, especially among older children and adults, is likely to generate symptoms of lactose intolerance in China, and, to a more limited extent in India. It is striking how little reference there is to this issue in the discussion of the global ascendance of milk. Lactose intolerance is sometimes noted in articles about China's new "thirst for dairy" and stomach complaints are commonly described among those calling the customer service hotline at the Shanghai Bright dairy (Chen 2003a). The only reference in popular literature in India that I have come across was the statement in the Nandini milk ad that lactose intolerance is "rare" and hence of little consequence. In a context in which cows' milk is partly a national symbol, discussion of variation in its tolerance is inconvenient. The same is true in the U.S. where rates of lactose intolerance are highest in minority populations, and the dairy industry has been forced to recognize this "problem" and provide solutions to "overcome" it (cf. Wiley 2004; see also Chapter 2).

This challenge is met by deploying local culinary traditions. In India yogurt (*dahi*), which makes use of bacterial fermentation of lactose, is a widely consumed milk product, and is often made in a liquid form (*lassi*) that is now commercially available. In China, the rise in drinkable yogurt intake has been double that of the already spectacular increase in milk intake (Fuller et al. 2006). Yogurt's visual and tactile similarity to soymilk and tofu, both created from the fermentation of soybeans and widely consumed, may facilitate its acceptance. It is worth noting that in Japan and South Korea, fluid milk consumption has been declining since the mid-1990s, when per capita intake peaked at around 35 liters per year, well below that of the United States (Dong 2006; Schluep Campo and Beghin 2006; see also Table 5.1). It may be that in Asia there is a lower ceiling for fluid milk consumption, related in part to high rates of lactose intolerance.

As milk has become a globalized commodity, produced and consumed in countries that heretofore had not had substantial dairy industries, it is safe to say that the "lactophobes" no longer fear milk, while consumption among the traditional "lactophiles" continues to slide. In China, milk has become the quintessential modern Western drink, identified with politically and economically strong nations, athletic success, and the means to make up for past "growth deficits." In India, milk has a long history of cultural valuation, yet it is difficult to make use of traditional motifs; modern nutritional science overcomes this limitation by positioning milk as nutrient rich and hence of benefit to the poor and a science-literate middle class. Enhancements of growth and contributions to physical strength of individuals and the nation predominate. In the U.S. milk retains the status of a more traditional drink, and beliefs about its contribution to child growth remain but are not sufficient to sell more milk. Current efforts represent attempts to update milk's image as a "wellness" drink" suitable for health-conscious and aging adults, especially women, but milk's long-standing association with child growth (as opposed to weight loss or solutions to health problems of the aged) is proving resistant to change.

In sum, while historically positive and negative evaluations of milk mapped on to genetic variation in adult milk digestion, the promise of milk's benefits to one's body and society has proved sufficiently alluring, and traditional means of food processing (e.g., fermentation) provide a means to circumvent the gastrointestinal discomfort that may accompany fresh milk consumption. Milk's alliance with growth has proven key to the spread of milk consumption around the globe and has overridden historical biases against milk; yet the metaphor of growth has specific local meanings as well, and may no longer compel individuals in countries dealing with the consequences of "excess" growth. While India and China are the two largest countries in the world, and growth in relation to population size is resolutely out of fashion in each, enhancement of physical growth (India) or "population quality" (China) provides a viable alternative for the deployment of the growth metaphor by the dairy industry. With larger heights and overall body sizes indicative of individual prosperity and national ascendancy, milk's "natural" connection to growth has provided the mechanism by which milk consumption can meet these goals, and the rationale for both government promotion of milk and individual dietary choices.

6

CONCLUSION

In this book I have outlined ways in which milk has been imagined and re-imagined across human prehistory and history and across cultures. Humans have a relatively recent evolutionary history with other mammalian milks. The adoption of these by some human populations led to natural selection favoring alleles allowing for the digestion of milk (specifically the milk sugar lactose) throughout life. People were now free to imagine milk in a new way: a novel source of food that had once been the exclusive provenance of infants was now available to all age groups, and which could also be modified in myriad ways for various purposes: preservation, transport, cooking, etc. Not everyone jumped on the milk bandwagon though, and many human populations did not keep dairy animals or used them exclusively for dairy products rather than for fresh milk. For many of them fresh milk was imagined as undesirable at best or out-right disgusting.

As we have seen, cows' milk has now achieved a global presence, even among populations who disdained milk in the past and who do not produce lactase throughout life. People among these populations have embraced milk by imag-ining its powers to promote growth, thereby achieving personal and nationalist goals for size and strength. How these ideas articulate with experiences of lactose intolerance as more people have more experience with milk, and whether increased milk drinking will have the widely anticipated positive effects on height will be important topics for future investigation. That milk drinking should enhance growth seems intuitive, given that it is the exclusive food of infants, who grow faster than at any other life stage. Yet as we have seen, the effects of cows' milk on growth are not well established. Nonetheless, the view that children need milk to grow properly is firmly entrenched, having been established early in the 20th century by public health promoters and the dairy industry. While increases in height and milk intake paralleled each other, their paths have diverged, with height leveling off and milk consumption declining markedly to levels well below the early 20th-century surge in intake.

Milk's rise to prominence in the American diet appears to have been short-lived. Despite the widespread views that U.S. citizens should consume milk, less and less is drunk. With other drinks to choose from and adults not modeling milk-drinking behavior, children are consuming little milk and are no longer the primary target for milk advertisements. The dairy industry has had to re-imagine milk now as a food for adults, one that has properties to alleviate the burden of chronic disease facing aging baby boomers. With milk having long been associated with children this poses a challenge, but one that should not prove insurmountable. After all, milk has always been modified and fortified to suit whatever public health needs have been most pressing at a given time: pasteurization for hygiene, Vitamin D for rickets (and now for a variety of chronic diseases, including some cancers), fat reduction for cardiovascular disease and stroke, probiotics and fiber for constipation, sugar to increase milk's desirability to children, evaporation or condensation for storage, etc.

But perhaps most importantly, milk is rarely consumed as fresh fluid milk, despite all the advances embraced and then mandated by sanitary reformers to make this form of milk safe and readily available. Ironically, dairy consumption patterns are reverting to their original form: milk as a food rather than a drink, processed into cheese or cultured into yogurt. And you can hardly miss the large jars of whey protein powder on the shelves with other dietary supplements; whey was consumed in fluid form in the past or made into cheese. People are "rediscovering" these historic forms of milk production and consumption (not necessarily with the same intent) while eschewing the 19th-century innovations that brought them fresh milk. While much is made of the central position of "nature's perfect food" in the American diet, in fact this is best thought of as a brief historical interlude rather than a natural or long-standing state of affairs.

In another twist, there is a growing movement to expand the consumption of raw, unpasteurized milk in the U.S. Much like 19th-century milk certification proponents worried that some of milk's "vital" components might be destroyed during pasteurization, contemporary raw milk advocates likewise suggest that raw milk contains important substances that are absent from pasteurized milk. States establish their own rules for intra-state milk commerce, and the sale of unpasteurized milk is currently illegal in 23 states. In some states individuals get around this by owning shares of a cow and thereby acquire their milk legally (there is no "sale"); or, as in my own state, Indiana, unpasteurized milk can be sold legally as "pet food" but not for human consumption. Two aspects of this movement are relevant to our discussion here. One is that unpasteurized milk is imagined as a more "natural" form of milk, harkening back to consumption patterns before pasteurization.[1] Of course we know that this was never a widespread phenomenon beyond whatever small amount might have been consumed opportunistically within a rural household. Furthermore, humans were cooking long before the milking of mammals (Wrangham 2009), so there is no a priori reason to assume that fresh milk was typically consumed raw.

Second, pasteurization was developed as a solution to milk contaminated in the unsanitary conditions of peri-urban swill milk barns. Mass consumption of milk was entirely predicated on pasteurization and refrigeration (although recall that the certification of raw milk was considered as an alternative but was judged to be too expensive). The Food and Drug Administration, which vehemently opposes the consumption of raw milk on public health grounds, views this trend using as a model the industrial-scale milk industry as the dominant milk-production system. For a local milk system, however, with small-scale production (where cows are not kept in large numbers and/or in confined spaces where infections spread), scrupulous attention to hygiene, and proximity of consumers who know how to handle raw milk (keep cold and use quickly!), raw milk can be consumed without detriment. That said, there have been some 1500 documented illnesses reported from drinking raw milk between 1993 and 2006, and two deaths (Centers for Disease Control and Prevention 2010). This number should be interpreted in light of the many more food-borne illnesses from products that are sold legally. The Centers for Disease Control report 76 million illnesses, 325,000 hospitalizations, and 5000 deaths from food-borne illnesses in the United States each year (Centers for Disease Control and Prevention 2010).

Milk pasteurization was also hailed as a solution to the ills of poverty that plagued many urban residents during the rapidly industrializing 19th century. Rather than confronting infectious diseases and undernutrition at their root source—poverty—a more tractable solution was to be found in a safe milk supply. Safe milk was both a source of nutrients and pathogen-free; as such it became a panacea for the more entrenched social problem of grinding urban impoverishment. It is not hard to see a contemporary parallel in National Dairy Council claims that health disparities in the U.S. can be—in part—attributable to the lower milk intake of minorities. Generally speaking, minorities are more likely to be poor and suffer disproportionately from many chronic diseases, and yes, they tend to drink less milk than European-Americans (Jarvis and Miller 2002). However, the solution to their problems is unlikely to be found in a glass of milk (or cheese, or yogurt), however convenient such a story might be for dairy promoters. While some milk nutrients such as calcium or Vitamin D (again, it is worth reminding ourselves that Vitamin D is not a natural constituent of milk!) may be involved in chronic disease risk, they—or the dairy foods they might come from—likely play only a small part in relation to other risk factors. The culinary traditions of minorities, which don't happen to include milk, have been disparaged much as those of Southern European immigrants were in the first half of the twentieth century (cf. Gabaccia 1998); of course the Mediterranean diet is now hailed as one of the most healthful!

Finally, once imagined as "the white poison," milk has become the marker of a healthy diet. Indeed, myriad publications and public health statements depict dairy-free diets as worrisome. In other words, it has become dangerous to not drink milk. The National Institute of Child Health and Development (NICHD) maintains its Milk Matters website precisely because this national health organization is concerned about calcium deficits in U.S. children. As I finished writing this book, the National Institutes of Health convened a Consensus Panel to discuss the lactose intolerance as a public health problem (February 22–24, 2010). From what has been made publicly available they concluded that:[1]

• Lactose intolerance is a real and important clinical syndrome, but its true prevalence is not known.

• The majority of people with lactose malabsorption do not have clinical lactose intolerance. Many individuals who think they are lactose intolerant are not lactose malabsorbers.

• Many individuals with real or perceived lactose intolerance avoid dairy products and ingest inadequate amounts of calcium and vitamin D, which may predispose them to decreased bone accrual, osteoporosis, and other adverse health outcomes. In most cases, individuals do not need to eliminate dairy consumption completely.

• Evidence-based dietary approaches with and without dairy foods and supplementation strategies are needed to ensure the appropriate consumption of calcium and other nutrients in lactose-intolerant individuals.

• Educational programs and behavioral approaches for individuals and their healthcare providers should be developed and validated to improve the nutrition and symptoms of individuals with lactose intolerance and dairy avoidance.

These conclusions are remarkably in line with dairy industry rhetoric, but importantly, dairy avoidance is itself portrayed as a problematic condition that stems from lactose intolerance (whether objectively assessed or perceived). The problems associated with dairy avoidance are framed entirely in terms of calcium and Vitamin D. However, the panel did note that "despite the widespread belief that decreased vitamin D and calcium intake associated with restricted intake of dairy products will lead to poor health outcomes, particularly related to bone mineral density and risk for fractures, *few data are available on bone health in individuals with lactose intolerance and dairy avoidance*" (emphasis added). Thus worries about the latter two "conditions" seem premature, given the current state of the science.

As I have shown in this book, dairy consumption is evolutionarily novel, historically rare across populations (most of East and Southeast Asia, Oceania and the whole New World), and routine milk drinking is only a little over 100 years old, suggesting that such worries may be misplaced. It is also worth bearing in mind that lactase impersistence is the ancestral condition for humans; it is not a "new" or "deviant" condition. It is only in a cultural context that privileges milk and dairy products that this ancestral biology becomes so. That said, in a context in which people lead sedentary lives, eat relatively few other sources of calcium (perhaps in part because they have been led to believe that milk is the sole source), and live in high-latitude environments where the ability to synthesize Vitamin D is limited for a major portion of the year, milk may very well be seen as a necessity in the diet, particularly in relation to bone health. But it is wise to step back and consider the larger picture into which milk fits, and the kinds of assumptions that are easy to make without taking everything into account. It is possible to imagine a post-weaning diet without milk or other dairy products—and the individuals who consume that diet—as not inherently "deficient."

NOTES

1 INTRODUCTION

1 While cows' milk is the most commonly produced and consumed milk, other mammals are milked by different human populations. As we will see, in India, the country that currently produces the largest amount of milk, water buffalo are the main source of milk. Other commonly milked mammals include goats, sheep, camels, and horses.

2 Although milk is a fluid and hence technically a "drink," I will refer to it as a food, which encompasses any substance that is ingested and contains nutrients. On occasion I will consider the variety of ways in which milk is consumed, which will include "solid" dairy products such as cheese or yogurt, but in this book the focus will be on fluid milk.

3 The term "milk" is often used to describe any opaque whitish fluid, whether of plant, animal, or synthetic origin, but it is likely that this more generic term stems from analogies with the material that mammalian mothers produce for their nursing young.

2 POPULATION VARIATION IN MILK DIGESTION AND DIETARY POLICY

1 Among Europeans, individuals with the C nucleotide at a locus in the 13th intron or a G in the 9th intron (these two introns being eight kilobases apart) within a DNA region about 14 kilobases upstream from the lactase gene produce lactase in adulthood while those with T or A at these same loci do not (Enattah et al. 2002). Among East African pastoralist populations, Tishkoff and colleagues (2007) identified three different single nucleotide polymorphisms (SNPs; G/C-14010, T/G-13915, and C/G-13907) that were associated with adult lactase production. Thus different mutations that allow for adult lactase production arose and were selected for in different dairying populations. Although the exact mechanism by which these loci regulate lactase production remains unclear, current evidence points to their action at the level of gene transcription, as most studies demonstrate variation in mRNA levels between those who do/do not continue to produce lactase, a pattern that becomes evident during childhood (Wang et al. 1998).

2 Every individual has two copies of each gene—one from the mother and one from the father. Different versions of a gene are called alleles. One allele can be dominant to the other, meaning it masks the expression of the other. In the case of lactase, the allele that causes lactase to remain turned on throughout life masks the expression of the one that turns it off; hence only one copy of the dominant allele will lead to lifelong lactase activity.

3 Cattle are not commonly kept in most sub-Saharan regions. In large part this is due to the widespread presence of tsetse flies, which are found in forested and woodland areas and along streams and by lakes. These flies are the vector of sleeping sickness (caused by the protozoan *Trypanosoma brucei*), a disease that is quite deleterious to cattle. Thus cattle herding is only found in more arid areas of sub-Saharan Africa, and lactase persistence is likewise rare in most parts of the region except for those where people herd cattle.

4　The National Dairy Council is now the education, communications, and nutrition research arm of Dairy Management Inc., which also includes the American Dairy Association and the U.S. Dairy Export Council.

5　Heaney and Weaver (1990) found that the calcium available in kale was actually higher than that in milk. Other sources disparage the calcium density and bio-availability of vegetable foods: "A person would need to consume 8 cups of spinach, nearly 5 cups of red beans, or 2¼ cups of broccoli to get the same amount of calcium absorbed from 1 cup of milk" (NDC 2003b).

6　This document may be accessed both through the ADA websites (www.eatright.org and www.webrd.org), or the National Dairy Council site (www.nationaldairycouncil.org), and is used in local dairy council websites as well.

7　These population groupings are regularly used in descriptions of variation in lactase activity in the United States. These are diverse kinds of groups, which reflect commonly used national, regional, "racial" or ethnic identities.

8　These groups have been described as "the most immediately dangerous ... because of their sometimes terrorist tactics" (Heaney 2001: 160). Further, their influence is decried: "We confront a recent, very modern efflorescence of militant groups that oppose all use of animal products and aim to effect a nutritional policy outcome similar to that of the creationists with regard to evolution. Those who care about nutrition, those who think nutrition important for the public health in general, need to realize that the present-day skirmishes may be only the first wave of a growing battle" (Heaney 2001: 163).

9　As we will see in Chapter 5, pro-milk attitudes and governmental policies are increasingly evident on a global scale. Thus immigrants to the United States may already come with attitudes shaped by these policies from their home countries.

10　In the popular "got milk?" ads, a variety of celebrities such as Whoopi Goldberg have worn lactose-reduced milk mustaches.

3　A BRIEF HISTORY OF MILK CONSUMPTION

1　Currently two forms of pasteurization are commonly used. One is "high-temperature short-time" (HTST). Milk is passed through warming pipes to raise the temperature to 161°F, where it is held for 15 seconds, then rapidly cooled. Ultra-high temperature (UHT) pasteurization raises the milk to a much higher temperature, but holds it there for only one to two seconds. It is often packaged in aseptic cartons, and has a long shelf life without refrigeration.

2　Coke was made available to soldiers during World War II and it was exempt from the rationing of sugar. It became an important symbol of American nationalism (Mintz 1996), and its rise in popularity among World War II troops may have contributed to the demise of milk in the post-World War II period.

4　MILK CONSUMPTION, CALCIUM, AND CHILD GROWTH

1　See: http://www.nichd.nih.gov/milk/milk.cfm

5　GROWING CHILDREN AROUND THE WORLD

1　www.parleproducts.com/brands/biscuits_milkshakti.asp

2　Indeed the 2008 melamine scandal in China, which was but one in a series of scandals involving contaminated milk and milk-based baby formula, caused domestic milk sales to plummet. But the cost of imported milk was prohibitive to many, and consumption of domestic milk is now up to 80% of its previous levels (Huei 2009). Industry analysts have described this as a "speedy recovery beyond anyone's expectations" but more than anything else it reveals how well established milk has become as part of the daily diet of many middle-class Chinese citizens.

3 This promotion has already given way to the 2010 "Building Strong Families" theme. In this new marketing scheme, milk is no longer emphasized for middle-aged women, but rather for the whole family, with benefits outlined—in general form—for each age group. See www.whymilk.com, which regularly changes its marketing.

6   CONCLUSION

1 http://consensus.nih.gov/2010/lactosestatement.htm

REFERENCES

Adamson, Melitta Weiss 2004 Baby-food in the Middle Ages. In *Nurture: Proceedings of the Oxford Symposium on Food and Cookery 2003*, R. Hosking, ed., pp. 1–11. Bristol, UK: Footwork.

Alcock, Joan P. 2004 "Please, Sir, I want some more:" Feeding of children in workhouses in the 19th century. In *Nurture: Proceedings of the Oxford Symposium on Food and Cookery 2003*, R. Hosking, ed., pp. 12–25. Bristol, UK: Footwork.

Ali, Jabir 2007 Structural changes in food consumption and nutritional intake from livestock products in India. *South Asia Research* 27(2): 137–151.

Aloia, John F. 2008 African Americans, 25-hydroxyvitamin D, and osteoporosis: a paradox. *American Journal of Clinical Nutrition* 88(2): 545S–550S.

American Academy of Pediatrics 1985 "Inactive" ingredients in pharmaceutical products. *Pediatrics* 76(4): 635–643.

American College of Gastroenterology 2009a Common Gastrointestinal Problems: A consumer health guide. Food Intolerance http://www.acg.gi.org/patients/pdfs/CommonGI Problems3.pdf. Accessed June 30, 2009.

—— 2009b Understanding Food Allergies and Intolerances: Lactose Intolerance http://www.gastro.org/wmspage.cfm?parm1=5679#Lactose%20Intolerance. Accessed June 30, 2009.

Anderson, E.N. 1987 *The Food of China*. New Haven, CT: Yale University Press.

Anderson, John J.B. 2001 Calcium requirements during adolescence to maximize bone health. *Journal of the American College of Nutrition* 20(2): 186S–191S.

Arnold, Wayne 2007 A thirst for milk bred by new wealth sends prices soaring. In *New York Times* 4 September 2007. New York.

Atkins, P.J. 1992 White poison? The social consequences of milk consumption, 1850–1930. *Social History of Medicine* 5: 207–227.

Atkins, Peter, and Ian Bowler 2001 *Food in Society: Economy, culture, geography*. New York: Oxford University Press.

Baker, Ian A. et al. 1980 A randomised controlled trial of the effect of the provision of free school milk on the growth of children. *Journal of Epidemiology and Community Health* 34: 31–34.

Batra, S.M. 1986 The sacredness of the cow in India. *Social Compass* 33(2–3): 163–175.

Bauer, R.L. 1998 Ethnic differences in hip fracture: A reduced incidence in Mexican Americans. American *Journal of Epidemiology* 127: 145–149.

Baumes, Harry 2004 China's Gowing Thirst for Dairy www.globalinsight.com/Perspective/PerspectiveDetail695.htm. Accessed June 21, 2010.

Bayless, T.M., and N.S. Rosensweig 1966 A racial difference in incidence of lactase deficiency: A survey of milk intolerance and lactase deficiency in healthy adult males. *Journal of the American Medical Association* 197: 968–972.

—— 1967 Incidence and implications of lactase deficiency and milk intolerance in white and Negro populations. *Johns Hopkins Medical Journal* 121: 54–64.

Beghin, John C. 2006 Evolving dairy markets in Asia: Recent findings and implications. *Food Policy* 31: 195–200.

Beja-Pereira, A. et al. 2003 Gene-culture coevolution between cattle milk protein genes and human lactase genes. *National Genetics* 35(4): 311–313.

Bekaert, Geert 1991 Caloric consumption in industrializing Belgium. *Journal of Economic History* 51(3): 633–655.

Berkey, Catherine S. et al. 2009 Dairy consumption and female height growth: prospective cohort study. *Cancer Epidemiology, Biomarkers, and Prevention* 18(6): 1881–1887.

Bernard, R.-J. 1975 Peasant diet in eighteenth-century Gevaudan. In *European Diet from Pre-Industrial to Modern Times*, E. Forster and R. Forster, eds, pp. 19–46. New York: Harper & Row.

Bertron, Patricia, Neal D Barnard, and Milton Mills 1999 Racial bias in federal nutrition policy, Part I: The public health implications of variations in lactase persistence. *Journal of the National Medical Association* 91(3): 151–157.

Black, Ruth E. et al. 2002 Children who avoid drinking cow milk have low dietary calcium intakes and poor bone health. *American Journal of Clinical Nutrition* 76: 675–680.

Bogin, Barry 1998 The tall and the short of it. *Discover* 19(February): 40–44.

Bonjour, J-P. et al. 2001 Gain in bone mineral mass in prepubertal girls 3.5 years after discontinuation of calcium supplementation: A follow-up study. *Lancet* 358: 1208–1212.

—— 1997 Calcium-enriched foods and bone mass growth in prepubertal girls: Randomized, double-blind, placebo-controlled trial. *Journal of Clinical Investigation* 99: 1287–1294.

British Broadcasting Corporation 2007 China drinks its milk. In BBC News 7 August 2007.

Brownell, Susan 2005 Food, hunger, and the state. In *The Cultural Politics of Food and Eating*, J.L. Watson and M.L. Caldwell, eds, pp. 251–258. Malden, MA: Blackwell.

Burger, J. et al. 2007 Absence of the lactase-persistence-associated allele in early Neolithic Europeans. *Proceedings of the National Academy of Sciences* 104(10): 3736–3741.

Byers, Katherine G. and Dennis A. Savaiano 2005 The myth of increased lactose intolerance in African-Americans. *Journal of the American College of Nutrition* 24(suppl. 6): 569S–573S.

Cadogan, Joanna et al. 1997 Milk intake and bone mineral acquisition in adolescent girls: Randomised, controlled intervention trial. *British Medical Journal* 315: 1255–1260.

Cameron, M.A. et al. 2004 The effect of calcium supplementation on bone density in premenarcheal females: a co-twin approach. *Journal of Clinical Endocrinology and Metabolism* 89(10): 4916–4922.

Cameron, Noel 2002 *Human Growth and Development.* New York: Academic Press.

Centers for Disease Control and Prevention 2010 National Center for Zoonotic, Vector-Borne, and Enteric Diseases http://www.cdc.gov/nczved/divisions/dfbmd/diseases/#f. Accessed May 17, 2010.

Chan, Gary M., Karen Hoffman, and Martha McMurry 1995 Effects of dairy products on bone and body composition in pubertal girls. *Journal of Pediatrics* 126: 551–556.

Cheer, Susan M., and John S. Allen 1997 Lactose digestion capacity and perceived symptomatic response after dairy product consumption in Tokelau Island migrants. *American Journal of Human Biology* 9(2): 233–246.

Chen, Kathy 2003a Got Milk? The new craze in China is dairy drinks http://www.mindfully. org/Food/2003/China-Dairy-Drinks28feb03.htm. February 28, 2003. Accessed September 20, 2005.

—— 2003b Dairy Firms Churn Out Milk Products in China http://www.tschang.net/ articles/20030301.htm. February 28, 2003. Accessed April 20, 2006.

Chiapparino, Francesco 1995 Milk and fondant chocolate and the emergence of the Swiss chocolate industry at the turn of the twentieth century. In *Food and Mateiral Culture.* M.R. Scharer and A. Fenton, eds, pp. 328–344. Edinburgh: Tuckwell Press.

Clark, Gregory, Michael Huberman, and Peter H. Lindert 1995 A British food puzzle. *The Economic History Review* 48(2): 215–237.

Cohen, Robert 1997 *Milk: The Deadly Poison.* Englewood Cliffs, NJ: Argus Publishing.

Comaroff, John, and Jean Comaroff 1992 *Ethnography and the Historical Imagination.* Boulder, CO: Westview Press.

Cook, Judith et al. 1979 The influence of availability of free school milk on the height of children in England and Scotland. *Journal of Epidemiology and Community Health* 33: 171–176.

Copley, M. S. et al. 2003 Direct chemical evidence for widespread dairying in prehistoric Britain. *Proceedings of the National Academy of Sciences of the United States of America* 100(4): 1524–1529.

Cordain, Loren et al. 2005 Origins and evolution of the western diet: Health implications for the 21st century. *American Journal of Clinical Nutrition* 81: 341–354.

Craig, Lee A., Barry Goodwin, and Thomas Grennes 2004 The effect of mechanical refrigeration on nutrition in the United States. *Social Science History* 28(2): 325–336.

Craig, O.E. et al. 2005 Did the first farmers of central and eastern Europe produce dairy foods? *Antiquity* 79(306): 882–894.

Cranney, Ann et al. 2008 Summary of evidence-based review on Vitamin D efficacy and safety in relation to bone health. *American Journal of Clinical Nutrition* 88(2): 513S–519S.

Crosby, Alfred, W. 1986 *Ecological Imperialism: The biological expansion of Europe 900–1900.* New York: Cambridge University Press.

Crumbine, Samuel J., and James A. Tobey 1929 *The Most Nearly Perfect Food: The story of milk.* Baltimore: The Williams & Wilkins Company.

Cullen, Louis Michael 1992 Comparative aspects of Irish diet, 1550–1850. In *European Food History: A Research Review.* H.J. Teuteberg, eds, pp. 45–55. New York: Leicester University Press.

Dairy Industries International 2006 Chinese dairies made their mark at the IDF World Dairy Summit http://www.dairyindustries.com/story.asp?id=2024211. Accessed July 9, 2007.

de Onis, M. et al. 2007a Comparison of the WHO child growth standards and the CDC 2000 growth charts. *Journal of Nutrition* 137(1): 144–148.

—— 2007b Development of a WHO growth reference for school-aged children and adolescents. *Bulletin of the World Health Organization* 85(9): 660–667.

Deaton, Angus 2008 Height, health, and inequality: The distribution of adult heights in India. *American Economics Review* 98(2): 468–474.

Delgado, Christopher F., and Clare A. Narrod 2002 *Impact of Changing Market Forces and Policies on Structural Change in the Livestock Industries of Selected Fast-Growing Developing Countries. Final Research Report of Phase I - Project on Livestock Industrialization, Trade and Social-Health-Environment Impacts in Developing Countries.* Rome: Food and Agricultural Organization.

Delgado, Christopher L. 2003 Rising consumption of meat and milk in developing countries has created a new food revolution. *Journal of Nutrition* 133: 3907S–3910S.

den Hartog, Adel P. 1992 Modern nutritional problems and historical nutrition research, with special reference to the Netherlands. In *European Food History: A Research Review.* H.J. Teuteberg, ed., pp. 56–70. New York: Leicester University Press.

—— 2001 Changing perceptions on milk as a drink in western Europe: The case of the Netherlands. In *Drinking: Anthropological Approaches.* I. de Garine and V. de Garine, eds, pp. 96–107. New York: Berghahn Books.

Dibba, Bakary et al. 2000 Effect of calcium supplementation on bone mineral accretion in Gambian children accustomed to a low-calcium diet. *American Journal of Clinical Nutrition* 71: 544–549.

Dong, Fengxia 2006 The outlook for Asian dairy markets: The role of demographics, income, and prices. *Food Policy* 31: 260–271.

Du, Xueqin et al. 2004 School-milk intervention trial enhances growth and bone mineral accretion in Chinese girls aged 10–12 years in Beijing. *British Journal of Nutrition* 92(1): 159–168.

Dupras, Tosha L., Henry P. Schwarcz, and Scott I Fairgrieve 2001 Infant feeding and weaning practices in Roman Egypt. *American Journal of Physical Anthropology* 115: 204–212.

DuPuis, E. Melanie 2002 *Nature's Perfect Food: How milk became America's drink*. New York: New York University Press.

Durham, William 1991 *Coevolution: Genes, culture and human diversity*. Stanford, CA: Stanford University Press.

Eaton, S. Boyd, S. B. Eaton III, and Melvin J. Konner 1999 Paleolithic nutrition revisited. In *Evolutionary Medicine*. W.R. Trevathan, E.O. Smith, and J.J. McKenna, eds, pp. 313–332. New York: Oxford University Press.

Eden, Trudy 2006 *Cooking in America, 1590–1840*. Westport, CT: Greenwood Press.

Enattah, Nabil Sabri et al. 2002 Identification of a variant associated with adult-type hypolactasia. *Nature Genetics* 30: 233–237.

Epstein, Samuel S. 1996 Unlabeled milk from cows treated with biosynthetic growth hormones: A case of regulatory abdication. *International Journal of Health Services: Planning, Administration, Evaluation* 26(1): 173–185.

Ettinger, B. et al. 1997 Racial differences in bone density between young adult Black and White subjects persist after adjustment for anthropometric, lifestyle, and biochemical differences. *Journal of Clinical Endocrinology and Metabolism* 82(2): 429–434.

EuroFIR n.d. Composition of Dairy http://www.eurofir.net/printpage.asp?id=6196.

Evershed, R. P. et al. 2008 Earliest date for milk use in the Near East and southeastern Europe linked to cattle herding. *Nature* 455(7212): 528–531.

FAO 2008 Food Outlook Global Market Analysis. http://www.fao.org/docrep/010/ai466e/ai466e00.htm. Accessed August 29, 2010.

Fenton, Alexander 1992 Milk products in the everyday diet of Scotland. In *Milk and Milk Products from Medieval to Modern Times*. P. Lysaght, ed., pp. 41–47. Edinburgh: Canongate Press.

Feskanich, D. et al. 1997 Milk, dietary calcium, and bone fractures in women: A 12-year prospective study. *American Journal of Public Health* 87: 992–997.

Flatz, G. and H. W. Rotthauwe 1973 Lactose nutrition and natural selection. *Lancet* 2(7820): 76–77.

Food and Agriculture Organization 1998 Noting decline in milk consumption among children, International conference calls for increased milk advertising aimed at school children http://www.fao.org/WAICENT/OIS/PRESS_NE/PRESSENG/1998/pren9863.htm. Accessed May 10, 2010.

——— 2002 *Human Vitamin and Mineral Requirements*. Rome: Food and Nutrition Division, FAO.

Food and Agriculture Organization, and World Health Organization 1998 *Vitamin and mineral requirements in human nutrition: Report of a joint FAO/WHO expert consultation*. Geneva: World Health Organization.

Fox, Dov 2005 Human growth hormone and the measure of man. *The New Atlantis* 7(Fall 2004/Winter 2005): 75–87.

Frassetto, L. et al. 2000 Worldwide incidence of hip fracture. *Journal of Gerontology Series A: Biological Science and Medical Sciences* 55: M585–M592.

Freidberg, Susanne 2009 *Fresh: A perishable history*. Cambridge, MA: Harvard University Press.

French, Simone A., Jayne A. Fulkerson, and Mary Story 2000 Increasing weight-bearing physical activity and calcium intake for bone mass growth in children and adolescents: A review of intervention trials. *Preventive Medicine* 31: 722–731.

Fuller, Frank H. et al. 2004 *China's Dairy Market: Consumer demand survey and supply characteristics*. Ames, IA: Center for Agricultural and Rural Development, Iowa State University.

——— 2006 Got milk? The rapid rise of China's dairy sector and its future prospects. *Food Policy* 31: 201–215.

Gabaccia, Donna 1998 *We Are What We Eat: Ethnic food and the making of Americans*. Cambridge, MA: Harvard University Press.

Garnett, S. et al. 1999 Effects of gender, body composition and birth size on IGF-I in 7- and 8-year-old children. *Hormone Research* 52(5): 221–229.

Ge, K., J. Jia, and H. Liu 2007 Food-based dietary guidelines in China – practices and problems. *Annals of Nutrition and Metabolism* 51(Suppl. 2): 26–31.

Gibbons, M.J. et al. 2004 The effects of a high calcium dairy food on bone health in pre-pubertal children in New Zealand. *Asia Pacific Journal of Clinical Nutrition* 13(4): 341–347.

Gillette, Maris Boyd 2005 Children's food and Islamic dietary restrictions in Xi'an. In *The Cultural Politics of Food and Eating*. J.L. Watson and M.L. Caldwell, eds, pp. 106–121. Malden, MA: Blackwell.

Goldberg, Jeanne P., Sara C Folta, and Aviva Must 2002 Milk: Can a "good" food be so bad? *Pediatrics* 110(4): 826–832.

Gould, Stephen Jay 1981 *The Mismeasure of Man*. New York: W.W. Norton.

Gould, William 2004 *Hindu Nationalism and the Language of Politics in Late Colonial India*. New York: Cambridge University Press.

Greer, Frank R., Nancy F. Krebs, and MD Committee on Nutrition 2006 Optimizing bone health and calcium intakes of infants, children, and adolescents. *Pediatrics* 117: 578–585.

Griffin, Michael 2004 Issues in the development of school milk. In *School Milk Workshop, FAO Intergovernmental Group on Meat and Dairy Products*. Winnepeg, Canada: FAO. http://www.fao.org/es/esc/common/ecg/169/en/School_Milk_FAO_background.pdf. Accessed August 29, 2010.

Grillenberger, Monika et al. 2003 Food supplements have a positive impact on weight gain and the addition of animal source foods increases lean body mass of Kenyan schoolchildren. *Journal of Nutrition* 133: 3957S–3964S.

Gunnell, D. et al. 2003 Are diet–prostate cancer associations mediated by the IGF axis? A cross-sectional analysis of diet, IGF-I and IGFBP-3 in healthy middle-aged men. *British Journal of Cancer* 88(11): 1682–1686.

Gupta, Charu 2001 The icon of mother in late colonial North India: 'Bharat Mata', 'Matri Bhasha' and 'Gau Mata'. *Economic and Political Weekly* 36(45): 4291–4299.

Harris, Marvin 1985 *Good to Eat: Riddles of food and culture*. Prospect Heights, IL: Waveland Press.

Hartley, Robert M. 1977[1842] *An Historical, Scientific and Practical Essay on Milk as an Article of Human Sustenance*. New York: Arno Press.

Heaney, Robert P. 2000 Calcium, dairy products and osteoporosis. *Journal of the American College of Nutrition* 19(2): 83S–99S.

—— 2001 The dairy controversy: Facts, questions, and polemics. In *Nutritional Aspects of Osteoporosis*. P. Burckhardt, B. Dawson-Hughes, and R.P. Heaney, eds, pp. 155–164. New York: Academic Press.

Heaney, Robert P., and C.M. Weaver 1990 Calcium absorption from kale. *American Journal of Clinical Nutrition* 51: 656–657.

Hegsted, D. Mark 2001 Fractures, calcium, and the modern diet. *American Journal of Clinical Nutrition* 74: 571–573.

Heyman, Melvin B., and the Committee on Nutrition 2006 Lactose intolerance in infants, children, and adolescents. *Pediatrics* 118(3): 1279–1286.

Hoffpauir, Robert 1982 The water buffalo: India's other bovine. *Anthropos* 77(1–2): 215–238.

Holmes, Michelle D. et al. 2002 Dietary correlates of plasma insulin-like growth factor I and insulin-like growth factor binding protein 3 concentrations. *Cancer Epidemiology, Biomarkers, and Prevention* 11: 852–861.

Hoppe, Camilla et al. 2004 Animal protein intake, serum insulin-like growth factor I, and growth in healthy 2.5-y-old Danish children. *American Journal of Clinical Nutrition* 80: 447–452.

Huang, H.T. 2002 Hypolactasia and the Chinese diet. *Current Anthropology* 43(5): 809–819.

Huei, Pei Shing 2009 China milk back in favor. In *The Straits Times,* January 9. Singapore: Singapore Press Holdings.

Ingram, C. J. et al. 2007 A novel polymorphism associated with lactose tolerance in Africa: multiple causes for lactase persistence? *Human Genetics* 120(6): 779–788.

—— 2009 Lactose digestion and the evolutionary genetics of lactase persistence. *Human Genetics* 124(6): 579–591.

Jarvis, Judith K., and Gregory D. Miller 2002 Overcoming the barrier of lactose intolerance to reduce health disparities. *Journal of the National Medical Association* 94(2): 55–66.

Jing, Jun 2000 Introduction: Food, children, and social change in contemporary China. In *Feeding China's Little Emperors: Food, children, and social change.* J. Jing, ed., pp. 1–26. Stanford, CA: Stanford University Press.

Juskevich, Judith C., and C. Greg Guyer 1990 Bovine growth hormone: food safety evaluation. *Science* 249: 875–884.

Juul, Anders et al. 1994 Serum insulin-like growth factor-I in 1030 healthy children, adolescents, and adults: Relation to age, sex, stage of puberty, testicular size, and body mass index. *Journal of Clinical Endocrinology and Metabolism* 78: 744–752.

Kalkwarf, Heidi J., Jane C. Khoury, and Bruce P. Lanphear 2003 Milk intake during childhood and adolescence, adult bone density, and osteoporotic fractures in US women. *American Journal of Clinical Nutrition* 77: 257–265.

Katz, R.S. 1981 Dairy Council perspective on lactose intolerance. In *Lactose Digestion: Clinical and nutritional implications.* D.M. Paige and T.M. Bayless, eds, pp. 263–268. Baltimore, MD: Johns Hopkins University Press.

Kelly, Owen, Siobhan Cusack, and Kevin D Cashman 2003 The effect of bovine whey protein on ectopic bone formation in young growing rats. *British Journal of Nutrition* 90: 557–564.

Knips, Vivien n.d. Developing Countries and the Global Dairy Sector Part I Global Overview www.fao.org/ag/AGAInfo/projects/fr/pplpi/docarc/wp30.pdf. Accessed May 12, 2006.

Komlos, John, ed. 1994 *Stature, Living Standards, and Economic Development: Essays in anthropometric history.* Chicago, IL: University of Chicago Press.

Köpke, Nikola C.G. 2008 *Regional Differences and Temporal Development of the Nutritional Status in Europe from the 8th Century B.C. until the 18th Century A.D.* PhD Dissertation. Universität Tübingen.

Korom, Frank J. 2000 Holy Cow! The apotheosis of Zebu, or why the cow is sacred in Hinduism. *Asian Folklore Studies* 59(2): 181–203.

Lampl, Michelle, Francis E. Johnston, and Laurence A. Malcolm 1978 The effects of protein supplementation on the growth and skeletal maturation of New Guinean school children. *Annals of Human Biology* 5(3): 219–227.

Lanau, Amy Joy, and Neal D. Barnard 2008 Dairy and weight loss hypothesis: An evaluation of the clinical trials. *Nutrition Reviews* 66(5): 272–279.

Lanou, Amy Joy, Susan E. Berkow, and Neal D. Barnard 2005 Calcium, dairy products, and bone health in children and young adults: A reevaluation of the evidence. *Pediatrics* 115(3): 736–743.

Lau, E.M. et al. 1990 Hip fracture in Hong Kong and Britain. *International Journal of Epidemiology* 168: 905–911.

Lee, Guy Carleton 1900 *Leading Documents of English History.* H.L. Jones, trans. New York: Henry Holt.

Lee, Warren T.K. et al. 1994 Double-blind, controlled calcium supplementation and bone mineral accretion in children accustomed to a low-calcium diet. *American Journal of Clinical Nutrition* 60: 744–750.

—— 1995 A randomized, double-blind controlled calcium supplementation trial, and bone and height acquisition in children. *British Journal of Nutrition* 74: 125–139.

Leighton, Gerald, and Mabel L. Clark 1929 Milk consumption and the growth of school-children. *Lancet* 213(5520): 40–43.

Levenstein, Harvey A. 1988 *Revolution at the Table: The transformation of the American diet*. New York: Oxford University Press.

Little, Michael A., and Brooke R. Johnson 1987 Mixed-longitudinal growth of nomadic Turkana pastoralists. *Human Biology* 59(4): 695–707.

Little, Michael A., Kathleen Galvin, and Mutuma Mugambi 1983 Cross-sectional growth of nomadic Turkana pastoralists. *Human Biology* 55(4): 811–830.

Lu, Ethel 2009 Radical shifts in China's milk market. http://www.china.org.cn/business/news/2009–01/14/content_17105973.htm 14 January 2009.

Lysaght, Patricia, ed. 1992 *Milk and Milk Products from Medieval to Modern Times*. Edinburgh: Canongate Press.

Mahias, Marie-Claude 1988 Milk and its transmutations in Indian Society. *Food and Foodways* 2: 265–288.

Marlowe, Thomas J., and James A. Gaines 1958 The influence of age, sex, and season of birth of calf, and age of dam on preweaning growth rate and type score of beef calves. *Journal of Animal Science* 17(3): 706–713.

McCollum, Elmer V. 1957 *A History of Nutrition*. Boston, MA: Houghton Mifflin.

McDowell, Margaret A. et al. 2008 Anthropometric reference data for children and adults: United States, 2003–2006. *National Health Statistics Report No. 10*. Hyattsville, MD: National Center for Health Statistics.

Mendelson, Anne 2008 *Milk: The surprising story of milk through the ages*. New York: Alfred A. Knopf.

Merrilees, M.J. et al. 2000 Effects of dairy food supplements on bone mineral density in teenage girls. *European Journal of Nutrition* 39: 256–262.

Miech, R.A. et al. 2006 Trends in the association of poverty with overweight among US adolescents, 1971–2004. *Journal of the American Medical Association* 295(20): 2385–2393.

Milk Processor Education Program n.d. a Carson Daly http://whymilk.org/bios/carson_daly.html. Accessed September 20, 2005.

—— n.d. b Got Milk? Get Tall! http://www.whymilk.com/facts_gotmilk.htm September 8, 2006. Accessed September 20, 2006.

—— 2009 The New Face of Wellness http://www.whymilk.com/new_face_of_wellness.php. Accessed June 1, 2009.

Milligan, Lauren A., and R.P. Bazinet 2008 Evolutionary modifications of human milk composition: Evidence from long-chain polyunsaturated fatty acid composition of anthropoid milks. *Journal of Human Evolution* 55(6): 1086–1095.

Mintz, Sidney 1985 *Sweetness and Power: The place of sugar in modern history*. New York: Viking.

—— 1996 *Tasting Food, Tasting Freedom: Excursions into eating, culture, and the past*. Boston, MA: Beacon Press.

Mullaly, John 1853 *The Milk Trade of New York and Vicinity, Giving an Account of the Tale of Pure and Adulterated Milk*. New York: Fowlers and Wells.

National Center for Health Statistics 2005 National Health and Nutrition Examination Survey http://www.cdc.gov/nchs/about/major/nhanes/. Accessed September 10, 2009.

National Dairy Council 2003a Dairy Food Sensitivities: Facts and fallacies www.nationaldairycouncil.org/lvl04/nutrilib/digest/dairydigest_683b.htm. Accessed May 30, 2003

—— 2003b Lactose Intolerance and Minorities: The real story http://www.nationaldairycouncil.org/SiteCollectionDocuments/LI%20and%20Minorities_The%20Real%20%Story%202010.pdf. Accessed June 21, 2010.

National Institute of Nutrition 2006 Dietary Guidelines for Indians http://www.indg.gov.in/health/nutrition/dietary-guidelines-for-indians. Accessed April 16, 2009.

National Institutes of Health 1994 Optimal calcium intake. *Journal of the American Medical Association* 272(24): 1942–1948.

Nestle, Marion 2002 *Food Politics: How the food industry influences nutrition and health.* Berkeley: University of California Press.

Nicklas, Theresa A. 2003 Calcium intake trends and health consequences from childhood through adulthood. *Journal of the American College of Nutrition* 22: 340–356.

Norat, T. et al. 2007 Diet, serum insulin-like growth factor-I and IGF-binding protein-3 in European women. *European Journal of Clinical Nutrition* 61(1): 91–98.

Oftedal, Olav T., and Sara J. Iverson 1995 Comparative analysis of nonhuman milks. In *Handbook of Milk Composition.* R.G. Jensen, ed., pp. 749–789. New York: Academic Press.

Ogden, C.L. et al. 2006 Prevalence of overweight and obesity in the United States, 1999–2004. *Journal of the American Medical Association* 295(13): 1549–1555.

Opotowsky, Alexander P., and John P. Bilezikian 2003 Racial differences in the effect of early milk consumption on peak and postmenopausal bone mineral density. *Journal of Bone and Mineral Research* 18(11): 1978–1988.

Orland, Barbara 2004 Alpine milk: Dairy farming as a pre-modern strategy of land use. *Environment and History* 10: 327–364.

Oski, Frank A. 1977 *Don't Drink Your Milk!: The frightening new medical facts about the world's most overrated nutrient.* Chicago, IL: Wyden Books.

Paige, David M., T.M. Bayless, and George G. Graham 1972 Milk programs: Helpful or harmful to Negro children? *American Journal of Public Health* 62(11): 1486–1488.

Painter, James, Jee-Hyun Rah, and Yeon-Kyung Lee 2002 Comparison of international food guide pictorial representations. *Journal of the American Dietetic Association* 102(4): 483–489.

Patton, Stuart 2004 *Milk: Its remarkable contribution to human health and well-being.* New Brunswick, NJ: Transaction Publishers.

Pollan, Michael 2008 *In Defense of Food: An eater's manifesto.* New York: Penguin.

Popkin, Barry M. 2010 Patterns of beverage use across the lifecycle. *Physiology and Behavior* 100(1): 4–9.

Prentice, Ann 1996 Constituents of human milk. *Food and Nutrition Bulletin* 17(4) http://www. unu.edu/unupress/food/8f174e/8f174e04.htm. Accessed June 21, 2010.

Pritchard, Bill 2001 Current global trends in the dairy industry http://www.geosci.usyd.edu. au/users/pritchard/agrifood/dairy.pdf. Accessed August 30, 2010.

Reaney, Bernice C. 1922 *Milk and Our School Children.* Bureau of Education and Child Health Organization of America, ed. Washington DC: Government Printing Office.

Reynolds, W.L., T.M. DeRouen, and R.A. Bellows 1978 Relationships of milk yield of dam to early growth rate of straightbred and crossbred calves. *Journal of Animal Science* 47(3): 584–594.

Rich-Edwards, J.W. et al. 2007 Milk consumption and the prepubertal somatotropic axis. *Nutrition Journal* 6: 28.

Rifkind, Herbert R. 2007 Fresh Foods for the Army, 1775–1950 http://www.qmfound.com/ fresh_foods_for_the_army_1775_1950.htm. U.S. Army Quartermaster Foundation. Accessed February 3, 2010.

Riley, Gillian 2004 Images of infant nutrition: Sightings of food in group & child portraits. In *Nurture: Proceedings of the Oxford Symposium on Food and Cookery 2003.* R. Hosking, ed., pp. 228–240. Bristol: Footwork.

Rogers, I. et al. 2006 Milk as a food for growth? The insulin-like growth factors link. *Public Health Nutrition* 9(3): 359–363.

Rona, Roberto J., and Susan Chinn 1989 School meals, school milk and height of primary school children in England and Scotland in the eighties. *Journal of Epidemiology and Community Health* 43: 66–71.

Rosenzweig, Norton S. 1973 Lactose feeding and lactase deficiency. *American Journal of Clinical Nutrition* 26: 1166–1167.

Rozin, Paul 1983 Human food selection: The interaction of biology, culture and individual experience. In *The Psychobiology of Human Food Selection*. L.M. Barker, ed., pp. 225–254. Westport, CT: AVI Publishing.

—— 1990 Acquisition of stable food preferences. *Nutrition Reviews* 48: 106–113.

Sahi, T. 1994a Genetics and epidemiology of adult-type hypolactasia. *Scandinavian Journal of Gastroenterology* 29(Suppl 202): 7–20.

—— 1994b Hypolactasia and lactase persistence: Historical review and the terminology. *Scandinavian Journal of Gastroenterology* 29(Suppl 202): 1–6.

Sandgruber, Roman 1992 Nutrition in Austria in the industrial age. In *European Food History: A research review*. H.J. Teuteberg, ed., pp. 146–167. New York: Leicester University Press.

Sandler, Rivka Black et al. 1985 Postmenopausal bone density and milk consumption in childhood and adolescence. *American Journal of Clinical Nutrition* 42: 270–274.

Schluep Campo, Isabelle, and John C. Beghin 2006 Dairy food consumption, supply, and policy in Japan. *Food Policy* 31: 228–237.

Shammas, Carole 1990 *The Pre-Industrial Consumer in England and America.* Oxford: Clarendon Press.

Shetty, Prakash S. 2002 Nutrition transition in India. *Public Health Nutrition* 5(1A): 175–182.

Simoons, F.J. 1978 The geographic hypothesis and lactose malabsorption. *American Journal of Digestive Diseases* 23: 963–980.

—— 2001 Persistence of lactase activity among northern Europeans: A weighing of evidence for the calcium absorption hypothesis. *Ecology of Food and Nutrition* 40(5): 397–469.

Stearns, S.C. 1992 *The Evolution of Life Histories.* New York: Oxford University Press.

Steger, Manfred B. 2003 *Globalization: A very short introduction.* New York: Oxford University Press.

Suarez, F. L. et al. 1997 Tolerance to the daily ingestion of two cups of milk by individuals claiming lactose intolerance. *American Journal of Clinical Nutrition* 65: 1502–1506.

—— 1998 Lactose maldigestion is not an impediment to the intake of 1,500 mg calcium daily as dairy products. *American Journal of Clinical Nutrition* 68: 1118–1122.

Takahashi, Eiji 1966 Growth and environmental factors in Japan. *Human Biology* 38: 112–130.

—— 1971 Geographic distribution of human stature and environmental factors – an ecologic study. *Journal of the Anthropological Society of Nippon* (Jinruigaka Zasshi) 76: 259–285.

—— 1984 Secular trend in milk consumption and growth in Japan. *Human Biology* 56(3): 427–437.

Tandon, R. K. et al. 1981 Lactose intolerance in North and South Indians. *American Journal of Clinical Nutrition* 34(5): 943–946.

Teegarden, Dorothy et al. 1999 Previous milk consumption is associated with greater bone density in young women. *American Journal of Clinical Nutrition* 69: 1014–1017.

Teuteberg, Hans J. 1991 Food patterns in the European past. *Annals of Nutrition and Metabolism* 35: 181–190.

Tishkoff, S.A. et al. 2007 Convergent adaptation of human lactase persistence in Africa and Europe. *Nature Genetics* 39(1): 31–40.

Troy, Christopher S. et al. 2001 Genetic evidence for Near-Eastern origins of European cattle. *Nature* 410(6832): 1088–1091.

United States Dairy Export Council 2002 The world dairy trade outlook, 2001–1006. *World Dairy Markets* 12(1): 1–5.

United States Department of Agriculture 2002 *Report to Congress on the National Dairy Promotion and Research Program and the National Fluid Milk Processor Promotion Program.* Washington DC: United States Department of Agriculture.

—— 2008 Production, Supply, and Distribution Online.Foreign Agricultural Service.http://www.fas.usda.gov/psdonline/psdDownload.aspx Accessed May 20, 2009.

—— 2010a Fluid Milk Promotion Act of 1990 as amended through May 7, 2010. Agricultural Marketing Service.http://www.ams.usda.gov/AMSv1.0/getfile?dDocName=STELPRDC 5062171 Accessed August 30, 2010.

—— 2010b Nutrient Database for Standard Reference, Release 22. Agricultural Research Service.http://www.ars.usda.gov/Services/docs.htm?docid=8964. Accessed August 30, 2010.

United States Department of Commerce 2002 Statistical Abstract of the United States: 2002. Washington D.C.: United States Census.

United States Department of Health and Human Services and United States Department of Agriculture 2005 Dietary Guidelines for Americans.http://www.health.gov/dietaryguide-lines/dga2005/document/pdf/DGA2005.pdf. Accessed August 30, 2010.

USDA Economic Research Service 2002 The WIC Program: Background, Trends and Issues. USDA Economic Research Service.http://www.ers.usda.gov/publications/fanrr27/ Accessed May 10, 2010.

—— 2005 ERS Dairy Consumption Trends. USDA Economic Research Service. http://www.ers.usda.gov/Data/FoodConsumption/spreadsheets/foodloss/Dairy.xls#'Total fluidmilk'! A1. Accessed August 30, 2010.

USDA Food and Nutrition Service 2009 The National School Lunch Program: history and development. http://www.fns.usda.gov/cnd/lunch/AboutLunch/ProgramHistory. htm. Accessed August 30, 2010.

—— 2010 WIC Food Packages. USDA Food and Nutrition Services http://www.fns.usda.gov/ wic/benefitsandservices/foodpkg.HTM Accessed August 30, 2010.

USDA National Agriculture Library Food and Nutrition Information Center 2010 Ethnic/ Cultural Food Guide Pyramid.http://fnic.nal.usda.gov/nal_display/index.php?info_ center=4&tax_level=3&tax_subject=256&topic_id=1348&level3_id=5732 Accessed May 10, 2010.

van Winter, Johanna Maria 1992 The consumption of dairy products in the Netherlands in the fifteenth and sixteenth centuries. In *Milk and Milk Products from Medieval to Modern Times.* P. Lysaght, ed., pp. 3–13. Edinburgh: Canongate Press.

VandenBergh, Marjolein F.Q. et al. 1995 Physical activity, calcium intake, and bone mineral content in children in The Netherlands. *Journal of Epidemiology and Community Health* 49: 299–304.

Vicini, John et al. 2008 Survey of retail milk composition as affected by label claims regarding farm-management practices. *Journal of the American Dietetic Association* 108(7): 1198–1203.

Vigne, J.-D., and D. Helmer 2007 Was milk a "secondary product" in the Old World Neolithisation process? Its role in the domestication of cattle, sheep and goats. *Anthropozoologica* 42(2): 9–40.

Vijayapushpam, T. et al. 2003 A qualitative assessment of nutrition knowledge levels and dietary intake of schoolchildren in Hyderabad. *Public Health Nutrition* 6(7): 683–688.

Walker, Marcella D. et al. 2007 Determinants of bone mineral density in Chinese-American women. *Osteoporosis International* 18(4): 471–478.

—— 2008 Race and diet interactions in the acquisition, maintenance, and loss of bone. *Journal of Nutrition* 138(6): 1256S–1260S.

Wang, Yangxi et al. 1998 The genetically programmed down-regulation of lactase in children. *Gastroenterology* 114: 1230–1236.

Wardlaw, Gordon M., and Jeffrey S. Hampl 2007 *Perspectives in Nutrition.* New York: McGraw Hill.

Weingarten, Susan 2005 Children's foods in the Talmudic literature. In *Feast, Fast or Famine: Food and drink in Byzantium.* W. Mayer and S. Trzcionka, eds, pp. 147–160. Brisbane, Australia: Australian Association for Byzantium Studies.

Weinsier, Roland L., and Carlos L. Krumdieck 2000 Dairy foods and bone health: Examination of the evidence. *American Journal of Clinical Nutrition* 72: 681–689.

Wells, J.C. et al. 2008 Body shape in American and British adults: Between-country and inter-ethnic comparisons. *International Journal of Obesity* (London) 32(1): 152–159.

Whaley, Arthur L. 2003 Ethnicity/race, ethics, and epidemiology. *Journal of the National Medical Association* 95(8): 736–742.

White, Ann Folino 2009 Performing the promise of plenty in the USDA's 1933–34 World's Fair exhibits. *Text and Performance Quarterly* 29(1): 22–43.

WHO Multicentre Growth Reference Study Group 2006 *WHO Child Growth Standards: Length/height-for-age, weight-for-age, weight-for-length, weight-for-height and body mass index-for-age: Methods and development.* Volume 2009. Geneva, Switzerland: World Health Organization.

WHO/FAO 2004 *Vitamin and Mineral Requirements in Human Nutrition.* Geneva, Switzerland: World Health Organization.

Wiley, Andrea S. 2004 "Drink milk for fitness": The cultural politics of human biological variation and milk consumption in the United States. *American Anthropologist* 106(3): 506–517.

—— 2005 Does milk make children grow? Relationships between milk consumption and height in NHANES 1999–2002. *American Journal of Human Biology* 17(4): 425–441.

—— 2008 Cow's milk for "growth" across the lifecycle. Paper presented at the 2008 Annual Meeting of the American Anthropological Association.

—— 2009 Consumption of milk, but not other dairy products, is associated with height among U.S. preschool children in NHANES 1999–2002. *Annals of Human Biology* 36(2): 125–138.

—— 2010 Milk, but not dairy intake is positively associated with BMI among U.S. children in NHANES 1999–2004. *American Journal of Human Biology* 22: 517–525.

Wiley, Andrea S., and John S. Allen 2008 *Medical Anthropology: A biocultural approach.* New York: Oxford University Press.

Winzenberg, Tania et al. 2007 Calcium supplements in healthy children do not affect weight gain, height, or body composition. *Obesity* 15(7): 1789–1798.

World Health Organization 2003 *Diet, Nutrition and the Prevention of Chronic Diseases. Report of the joint WHO/FAO expert consultation.* Geneva, Switzerland: World Health Organization.

—— 2004 *Development of Food-based Dietary Guidelines for the Western Pacific region.* Geneva, Switzerland: World Health Organization.

—— 2007 *WHO AnthroPlus software.* Geneva, Switzerland.

Wrangham, Richard W. 2009 *Catching Fire: How cooking made us human.* New York: Basic Books.

Yang, X.G. et al. 2005 Study on weight and height of the Chinese people and the differences between 1992 and 2002. *Zhonghua Liu Xing Bing Xue Za Zhi* 26(7): 489–493.

Zhu, Kun et al. 2004 Bone mineral accretion and growth in Chinese adolescent girls following the withdrawal of school milk intervention: Preliminary results after two years. *Asia Pacific Journal of Clinical Nutrition* 13: S83.

INDEX

Note: Page numbers in **bold** are for figures and those in *italics* are for tables.